Embryogenesis
in Myth and
Science

Embryogenesis
in Myth and Science

THOMAS J. WEIHS

Floris
Books

First published in 1986 by Floris Books, Edinburgh
This third edition published in 2017
Second printing 2018

© Estate of Thomas Weihs 1986

British Library CIP Data available
ISBN 978-178250-499-3
Printed by Lightning Source

Contents

References

The system used in this book cites author and year of publication, followed by page. The full title and publication details are in the bibliography.

Acknowledgements

Figures 2, 4, 6, 8, 10 and 12 from *Old Testament Miniatures* by kind permission of Phaidon Press, Oxford. Figures 14 and 15 by kind permission of the British Library, London. Figures 17, 18, 19 and 20 from Jan Langman, *Medical Embryology*, by kind permission of William & Wilkins, Baltimore, and of Thomas Sadler, North Carolina. Figure 21 by kind permission of Schwabe & Co, Basle. Figure 22 by permission of Life Picture Service. Figure 23 by permission of ZEFA Picture Library (UK) Ltd.

Foreword

In his years as Medical Superintendent for Camphill, Thomas Weihs probably interviewed more than twelve thousand mentally handicapped children and adults. What sustains someone in such a work?

A pianist sits down at an old piano in a village hall. He plays the first movement of Beethoven's Moonlight Sonata. Some notes are missing from the piano and many strings are out of tune. Yet through the jangle, the sonata itself is unmistakable. We long to tune the piano as far as possible (we know it will never be a Steinway). And we wonder more than ever at the perfection of the sonata itself, however imperfectly embodied. We know it is the piano, not the music, which is 'handicapped'.

The fact that great numbers of people devote their lives to caring for the severely handicapped points to a tacit knowledge, apprehended in the heart rather than the head, a sense for the human individuality as an entelechy, a spiritual reality perfect in itself, however inadequate the instrument through which it has to sound. Asked to define such knowledge, to assure others of its certainty, we children of the scientific age are in difficulties. We retreat into talk about 'values'. Yet medical and biological technologies are now putting extraordinary new powers into our hands for modifying or repairing our bodies and for intervening in the earliest stages of our development. It is as though we were becoming responsible for a musical instrument business hitherto conducted by nature, without any understanding of music.

Earlier cultures knew rather little about the details of physical conception and gestation. They could intervene hardly at all. But they had very definite conceptions of the *meaning* of human life. These were much more substantial than mere 'values'. They were embodied in shared myths which shaped the whole of individual and community life. (We have only to look at the extraordinary richness and subtlety of the inner world of the Kalahari bushmen, unearthed by Laurens van der Post as he gained their trust. They were relatively unconcerned about losing their lives; but they knew that if they were to lose their 'stories', they would lose themselves.)

All such myths, however various in detail, share the same archetypal gesture: for our real origins they look to beings, to gods; and for our real home they look to an invisible, spiritual realm.

Where is a sonata 'at home'? Not in an old piano. Yet it can sound through even a jangled instrument, bringing with it, in Wordsworth's phrase, intimations of immortality. Aristotle, who stood at the threshold between a fading world of myth and a dawning world of physical facts, urged that in seeking to understand the cosmos, we must take account of four distinct kinds of causes: A potter makes a pot. Its *material cause* includes the properties of the clay (a potter can't use sawdust). Its shape emerges through the activities of the potter's hands – the *efficient cause*. But what guides the hands? The idea is in the potter's mind – the *formal cause*. And who determines what idea is embodied? The potter himself, the *final cause*.

Since Aristotle's time, the final cause, the beings who originate all things in myths, have faded completely from the scientific world. Only chance is allowed to haunt the scene. (So it is perhaps not surprising if we who are also beings wonder whether we really exist.) Formal causes linger on, ghostly and abstract, yet potent as what we call 'laws of nature'. We grasp them mathematically, but while we have discovered many formal causes for purely physical events, the true laws of biological form elude us, and will continue to be obscured as long as we continue to believe that they must be extensions of the laws of physics.

Likewise, efficient causes persist as 'energies' — mechanical, gravitational, electromagnetic, nuclear. They are physical and subphysical. We hardly know how to conceive of other energies — biological, psychological or spiritual — except by turning to older traditions, and risking heresy within scientific orthodoxy.

Thomas Weihs saw that if our 'values' are to survive in a

scientific age, the tacit, intuitive heart-knowledge which these embody, and which is the real source of meaning in every human life, needs to be raised into consciousness and married with science itself. This means taking myths seriously again as an inheritance of an older mode of insight, an insight into formal and final causes and the energies which bear them into the physical world. Yet not simply as a revival. How can *mythos* be reawakened as science, researched rather than inspired or remembered? It is a formidable undertaking.

He regarded this book, completed ten days before he died, as no more than the merest sketch of some possibilities. It is both fruit and seed from a life's work. In this, it is at one with his work with the handicapped. He was meeting and living with human beings whose physical situation, viewed in terms of material causes and chance, was often both 'senseless' and hopeless. But he sought to hear the music of their lives, and knew as heart-knowledge that he was attending one movement — problematic, perhaps, and deeply demanding, but neither the beginning nor the end of their symphony.

John Davy
(1927–1984)

Introduction

While this book is essentially about embryogenesis, it is also about mythos and science. Ultimately, it has been written out of my concern for our time. When, as an older person, one looks back over the present century, one can be deeply impressed by the progress in all technical, material, but also in social and care fields, by advances in communication and transport, in mechanization and automation, in medicine and agriculture, all of which have been phenomenal. Man has been relieved of and freed from countless burdens and threats, and standards of living have been widely raised. Probably all this progress is the direct fruit of science.

At the same time, crime and violence have increased; even medicine, once the art of healing, uses violent terminology; it has developed pain-*killers, anti*-biotics, sets out to *exterminate* diseases. It realizes the prevention of congenital pathologies by the destruction of the unborn or the starvation to death of the handicapped new-born.

World peace is maintained by the deterrent of nuclear weapons in quantities that could annihilate all human, animal and plant life on earth. All these terrifying prospects, too, are the fruits of science.

The argument put forward in this book is that this dilemma is inevitable if we continue to interpret science as the only source of truth and guidance. But this is what is largely happening. Agriculture, industry, commerce, medicine, education, social services, politics — are all based on the belief that they must be scientifically verified, although they have all originated in art, culture and religion. At the same time science has freed itself totally from ethics and aesthetics.

In this book, I have described how embryology has evolved

from myth to science and how modern scientific developments have brought to light new phenomena which could engender a new mythology. Yet would that not be a falling-back into a dark and unenlightened past? Is not the freeing of science from the bonds of ethics and aesthetics the very core of modern progress?

This certainly seems to be the case as long as we maintain that the law of causality is inherent in nature, is in fact, the essence of objective natural law. But 'law' strictly means that where there is *A*, *B* must follow. We cannot really mean that such a moral postulate is inherent in nature, yet this is just the origin of the concept of natural laws that, in the period from Newton to Darwin, scientists have held to be God-given or divine.

This is however, no longer the case in our time. Modern physics has shown that at sub-atomic and cosmic levels, 'objectivity' cannot be strictly maintained, that the results of experiments depend equally on the observer, that the one-directional flow from cause to effect cannot be absolutely upheld, that in place of law and necessity, we have to accept probability, that in place of the distinction between mass and energy, we must take into consideration process and becoming.

Another contribution made in our time which challenges the idea that causality is inherent in nature is K. Popper's concept of a threefold world: a world of things, a world of subjective experience (thinking, feeling and intention), and a third world of ideas and theories, of mathematics and logic. These three worlds interact, but none can be completely reduced to the other nor can any one exist without the others. While the third world of ideas and theories is largely the outcome of the second of subjective experience, it has its own independence. It seems that the idea of causality was maintained within this third realm, although it arises from the human activity of thinking in the second world. It certainly does not belong to the first, the world of things, the world of nature.

Science is undoubtedly not a natural phenomenon but a human one, born of human activity. It does not only obey the principles of objectivity and causality, but depends a great deal on imagination, inspiration and intuition. Darwin formulated his theory of evolution at a time when the generally accepted theory of heredity, held also by himself, made his theory of evolution practically impossible. He could not know that Mendel had already published his new theory of hereditary transmission, mysteriously forgotten for fifty years, to be rediscovered only at the beginning of the twentieth century. And yet, his genius or his intuition led Darwin to postulate his impossible theory, which was accepted by countless men, against all the evidence available at the time.

It is a well-known fact that in many cases, new scientific theories are less the result of experiment and observation, than of intuition, only later to be verified or at least rendered plausible by experiment. Several such instances have been referred to in this book.

At some point in what has been submitted here, it may appear as if I held that only some phenomena lend themselves to scientific solutions and others only to symbolical or mythical ones. This is not meant at all. While at one time, something may yield most fruitfully to the one approach, it may yield at another time or in another situation more fruitfully to the opposite one. What this book hopes to convey is that scientific theories, while divesting themselves of moral and ethical burdens and thus offering only tools, often open up new phenomena which can become the starting-points for fresh elucidations of ancient mythos or also initiate completely new symbolic and mythical insights. Thus science and mythos are not seen as contradictions, but as creative polarities, as male and female in nature.

I hope to have shown that just as the child is engendered between man and woman, so there can arise between the polarity of science and mythos, a sphere in which man is seen as a being in the becoming, as a promise, ever unfinished, ever open and hopeful.

The fundamental issue here however, is not to argue this point philosophically, but rather to have attempted to show that in some cases of scientific research, contributions have been made to a new symbolic and mythical understanding, to a new aesthetic and moral experience.

I am aware of the inadequacy of this endeavour. I am aware that some of the statements contained in this book will seem foolish or incomprehensible to some and erroneous to others. I do not wish to convince the reader of the validity of my arguments as they are essentially derived from my own subjective experience and musings. But I not only hope for forgiveness where I may have offended, but equally that some readers may feel stimulated to find their own way between scientific and mythical experience, thus creating for themselves and perhaps opening up for others, a space in which human beings can in freedom, dignity and love, grow and become.

In such a space, K. Popper's three worlds can appear in the garments of the Christian Trinity: World One of things and beings, of nature, as the World of the Father; World Two of subjective human experience in thinking, feeling and intention as the World of the Logos, the Christ; and World Three of ideas and absolutes as the World of the Holy Spirit.

Yet we must never forget that the Trinity is not only three separate independent entities, but a Triune. A process of oneness in which each part can be the whole, even the source and well-spring of the other two, is not only to be found in Popper's three Worlds, in the Tao – Yin and Yang – but probably in all phenomena of being.

The decision to write this book goes back to two sources: one is my encounter with Karl König, founder of the Camphill Communities, whom I heard giving a course of lectures on 'Embryology and Genesis' in Basle in 1948. Ever since, I have been captivated by the miracle through which out of one round cell, the highly differentiated human organism, with its millions of specialized cells, its organs and organ systems, and its ultimate form develops. Karl König had also introduced me to the work of Rudolf Steiner which continued to stimulate my thinking and my experience over the last forty years.

The second source is my deepening concern at the one-sided influence the natural sciences are exerting in our time, the most formidable example of which is probably that of modern physics. Sub-atomic physics, having completely transcended the mechanistic Newtonian world-concept and opened up unknown, most profound philosophical possibilities, has manifestly led to the threat of a nuclear holocaust. There are similar polarizations in the fields of other sciences.

This book is an attempt to show a way by which the one-sided use of science as the dangerous tool of control and power can be balanced by allowing the symbolic meaning of scientific discoveries to be heard. It embodies a desire to help to prepare in the human mind a space of freedom and dignity between science and mythos.

The first stepping-stone towards this book was perhaps Jonathan Stedall's filming of one of my talks on embryology to students in the Training Course at the Camphill Schools, *In defence of the Stork* in 1971. I want to record my thanks to Jonathan here.

That I did write this book at all is largely due to the sustaining help of my wife, Anke, and my thanks go to her as well as to my friend, Dr Hans Müller-Wiedemann, for his encouragement and support.

From Mythos
to Science

1. The story of creation

'In the beginning Nyx, who is Night, hovered in the darkness. An egg was laid by Nyx, the black-winged bird. From the upper shell of the egg was formed Ouranos, who is Heaven, and from the lower shell, Gaia, who is Earth. And Eros, who is Love, flew north from the egg. Drawn together by Eros, Ouranos and Gaia married, and they had for children the Titan Gods and Goddesses' and later, Kronos, who castrated his father, Uranos (Colum 1930, 61). From Kronos and Rhea were born the Olympian Gods, and ultimately Zeus, described in the Orphic fragments as: 'Zeus the First, Zeus the Last, the God with the blinding thunder. Zeus is the head, Zeus is the middle, from Zeus everything has its end. Zeus is the ground of the earth, and the starry sky. Zeus is the male, Zeus is the underlying woman. Zeus is the breath of all, Zeus is the soaring fire. Zeus is the root of the sea, Zeus the sun and moon. Zeus is the King, Zeus the beginning of all. . . .'

These words constitute a poetic synthesis of the Orphic Genesis and Hesiod's Theogeny, which is significant on two levels. Firstly, it expresses the oneness of two possible archetypes of creation — that of a unit (the egg) dividing into a polarity (Uranos-Gaia, heaven-earth) and giving rise to a middle third, Eros, who sets Heaven and Earth into motion, which is the Orphic conception — or secondly Hesiod's account in which, in the midst of Chaos, the yawning vacuum of nothing, there comes about the first finite reality, the middle ground of the earth, Gaia, separating the dark underworld, Erebos, from the light ether-filled heaven which is Uranos.

In a sense, we find the same element in the Chinese which sets out an archetypal ambiguity of primal beginning as Yin and Yang, arising as a polarization of the original Tao and yet, creating the Tao in their integrating harmony.

The other significant thing in Padraic Colum's synthesis is that in some manner, it expresses an essential feature in all Greek mythology, which seems to draw together the mythologies from all over the world and from all periods of civilization in a final fireworks of glory and majesty before setting out on the way from meaning to the causal, from mythos to science.

Not only in the Egyptian and Indian, but also in the Finno-Ugric and Polynesian parts of the world does one find the same basic creation lore with the egg as the origin, the separation of heaven and earth, and the emergence of a third principle from which, as in the first, the other two have arisen.

The lore of the Finno-Ugric peoples seems to have originated well before the separation of the two branches, one to the north to what is Finland today, the other to the plains of present-day Hungary, in the fourth millenium BC.

As related in the central Finnish epic, the *Kalevala*, creation is essentially the emergence of the earth from the seas. To begin with, there was water everywhere. A female eagle flies over the waters in search of a dry place to lay her egg. At last, she descries an island on the surface of the waters. She is not aware that it is the knee of Väinämöinen, the sorcerer, who is asleep in the waters. The eagle alights on what she takes to be land, lays her egg and begins to brood. The heat on his knee rouses Väinämöinen from his slumber. He jerks his knee and the egg falls into the water and breaks. From the yolk is formed the heaven with sun and moon; from the broken shell, the earth and the stars.

Among the Oceanic peoples, the Polynesians have an especially impressive creation myth. It tells of an absolute One, a continually metamorphosing chaos, or of a godhead, Io, who is the soul of the world and thus subject as well as object of creation at the same time. His breath is the creative chaos. From the lap of the creative chaos, the polarity of Papa and Rangi, the female Earth and the male Heaven, are engendered. Their union brings forth life, but life remains trapped in the embrace of the two and allows neither light nor darkness to emerge. The sons of Rangi and Papa confer as to how to release the captive life. Tu, later the god of war, proposes the destruction of Rangi and Papa, but Tan, the god of forests and birds, suggests their separation by non-violent means; this is supported by the other brothers with the exception of Tankiri, ruler over winds and storms. Tan then separates his parents and with his own body as pillar, holds up Rangi, the Heaven, above Papa, the Earth.

In the creation myth of the Maori, we likewise find a pre-existent egg, broken in two by Ta'aora and separated into Heaven and Earth.

The fertile and rich mythology of India with its countless gods and their unlimited attributes contains the following creation story: In the beginning the universe was shrouded in mist, unperceived, undistinguished, undiscovered, unknown, as if completely submerged in sleep. Then the Lord of All showed himself, irresistible, existing in himself, eternal essence of all Being. Wishing to bring forth from his body other creatures, he made the water and placed his seed into it. His seed became a golden Egg, Hiranyagarbha. It shone like the sun and he himself was born in it as Brahma, the ancestor of all worlds. After Brahma had dwelled for a year in the egg, he divided it in two by the power of his thought; the upper half of the shell became the heavenly or divine sphere of the universe, the lower half the material or earthly sphere. Between the two, the atmosphere, the earth floating on the waters and the cardinal points of the four directions took up their places. Out of the opening egg, the primeval being emerged with a thousand thighs, a thousand feet, a thousand arms, eyes, faces, a thousand heads. This primeval being or Purusha, sacrifices itself in order to create the world, whereby each of its members and parts brings forth whole series of creatures.

From the wealth of Egyptian mythology, we may take the creation story of Heliopolis: In the beginning, there was Nun, the limitless Chaos, a vast ocean of formless magma including within it the potential of life as well as the principle of consciousness, the god Atum, the Whole One, the Complete. To begin with, Atum is alone, and without the participation of a female principle, begets the seed from which a first pair of gods is engendered.

Shu and Tefnut — Shu, the male, representing the vastness and emptiness of the atmosphere, and Tefnut, the female, the life-giving moisture of the atmosphere. From the union of Shu and Tefnut, a second pair of gods springs forth — Geb, the god of the earth, and Nut, the goddess of the heaven. It is Shu, the god of the vastness of the atmosphere, who raises the body of Nut from her embrace with Geb so that she arches high above the earth as heaven. From Geb and Nut are born a further two pairs of gods — Osiris and Isis, and Seth and Nepthys. These complement but also contend with each other and would seem to represent the transition from a cosmic to an earthly order.

In the mythologies of the most diverse peoples all over the world, one can often discern three distinct phases in the respective creation stories: a first phase of beginning, creation stories such as those referred to above; a second phase which deals with generations of gods, titans, giants, sorcerers and their deeds; and finally a third phase telling of shamans, heroes, or of early kings, and then in our western civilization, of saints and knights, still within creation's mythology.

Later on, we shall find that contemporary human embryology distinguishes three phases of development not unlike the three phases of mythology: a first phase which precedes the appearance of the human form, a phase of general, archetypal creation; a second phase in which the human form emerges; and finally a third, the longest phase of development, differentiation, detail and growth.

Examples of creation myths have been cited here because in them, we may most easily recognize elements of early embryogenesis, which is the subject we wish to treat of.

It is usually maintained that the absoluteness of *monotheism* in Judaism, Christianity and Islam does not lend itself to the mythological, hence these religions are rarely referred to in textbooks on mythology. Modern Christian theology certainly tends to deny the quality of mythos in its teachings in favour of historicity and miracle.

In later chapters of this book, the suggestion will be made that

1. THE STORY OF CREATION

recent advances in biology can lead to a new understanding of Christianity as the mythos of modern man. But first we shall follow the path of the emergence and development of scientific thinking in the field of embryogenesis out of its mythological beginnings.

2. A brief history of embryology

Ancient Egyptian civilization which, as early as the dawn of the Old Kingdom in the third millenium BC had an elaborate school of medicine, does not seem to have considered embryogenesis, nor do early medical papyri make mention of the embryo. It is all the more remarkable that the ancient Egyptians, as well as the Chinese, mastered the art of the artificial incubation of birds' eggs long before Europeans carried out their first experiments in this field in the nineteenth century.

It seems that the skill of artificial incubation of eggs was lost only during the Middle Ages, for Pliny in the first Christian century gives a charming account of how Livia Augusta, Nero's wife, hatched a hen's egg in her bosom to find out whether she was carrying a boy or a girl. The egg hatched a cock and Livia Augusta gave birth to Tiberius Caesar.

Pliny likewise refers to the Egyptian practice of hatching eggs in large rooms heated with dung. Such incubators were still used in our own century in Egypt and also in China, temperatures being taken by an attendant who stayed in the-incubator, holding the blunt end of the eggs to his eyelids in lieu of a thermometer. In such incubating rooms, hundreds of thousands of eggs could be hatched annually with results that compare well with present-day scientific achievements in egg-incubation.

The ancient Egyptian cult of the placenta of which Needham (1934, 4) makes mention is also relevant to our concerns here. It would seem that for nearly three thousand years, up to the time of the Ptolemies in 300 BC, a standard representing the royal placenta was carried before the Pharaohs. During the Old Kingdom, a high royal office existed: The Opener of the King's Placenta, which symbolically signified the end of the king's reign, possibly his birth into death or the after-life that played so dominant a part in the religion of ancient Egypt. The placenta was referred to as the Bundle of Life, the Life Giver, the Sun, the Seat of the External Soul.

We have the beautiful hymn to the Sun God, Aton, written by that controversially heretic young king, Akhenaton (Amenophis IV) in 1400 BC (Needham 1934, 8f.):

> Creator of the germ in woman,
> Maker of the seed in man,
> Giving life to the son in the body of his mother,
> Soothing him that he may not weep,
> Nurse (even) in the womb.
> Giver of breath to animate every one that he maketh
> When he cometh forth from the womb on the day of his
>   birth.
>
> Thou openest his mouth in speech,
> Thou suppliest his necessity.
>
> When the fledgling in the egg chirps in the shell
> Thou givest him breath therein to preserve him alive.
> When thou hast brought him together
> To the point of bursting out of the egg,
> He cometh forth from the egg
> To chirp with all his might.
>
> He goeth about on his two feet
> When he hath come forth therefrom.

This touches upon the question as to the moment of animation in the embryo, a question which has continued to be asked throughout the ages, especially in the Middle Ages, and again more recently.

The relation of the placenta to the sun and the 'external soul' will be referred to in the further considerations of Chapter 7.

The Ayur-Vedas of the medical schools of ancient India, on the other hand, give remarkably detailed descriptions of embryonic development. They set forth four elements as the starting-point:

the father's semen, the mother's blood (more specifically the menstrual blood), both originating from *chyle* (the lymph fluid), as well as Atman the subtle body, and Manas the mind. Here the physical-organic elements and the subtle-spiritual elements together constitute the starting-point of embryonic development.

Further, according to the Ayur-Veda writings, the differentiation into head and limbs in the embryo begins in the third month. By the fourth month, thorax, abdomen and heart are formed, and in the sixth month, bones, sinews, veins as well as hair and nails appear. The hard parts of the body are derived from the father, the soft parts from the mother. Nourishment for the embryo was conducted through vessels bringing *chyle* from mother to foetus.

The idea that the actual early forming-processes in embryonic development could be likened to the clotting of milk by heat in the making of cheese appears in Indian embryology, as well as the idea of menstrual blood as the mother's initial contribution to embryonic development. These ideas were for a time widely upheld, and later propounded by Aristotle.

The latter idea may have contributed to the practice of child marriage in India in the belief that menstrual blood was the *materia prima* of the development of the embryo and that to lose any before marriage was paramount to infanticide.

A reference to the Greek Orphic cosmogonics of the eighth and seventh centuries BC, with which we opened the first chapter, really belongs here, for in them, the creation of the world is regarded as man's embryogenesis, a view which in a sense sustains this book. The Orphic cosmogonies are rather beautifully called up by Aristophanes in his comedy *The Birds* (414 BC) (Needham 1934, 10):

> Chaos was first, and Night, and the darkness of
> Emptiness, gloom tartarean, vast;
> Earth was not, nor Heaven, nor Air, but only the bosom
> of Darkness; and there with a stirring at last
> Of wings, though the wings were of darkness too, black
> Night was inspired a wind-egg to lay,

> And from that, with the turn of the seasons, there sprang
> to light the Desired,
> Love, and his wings were of gold, and his spirit as swift
> as the wind when it blows every way.
> Love moved in the Emptiness vast, Love mingled with
> Chaos, in spite of the darkness of Night,
> Engendering us, and he brought us at last to the light.

Fragments of writings handed down would indicate that pre-Socratic thinkers added various ideas to embryology, detailing times and sequences in foetal development, relating male and female elements to left and right sides, mammalian milk to egg white, some holding that the foetus fed through the mouth, others through the entire body like a sponge, among other interesting but fanciful ideas.

The school of medicine founded by Hippocrates (460–357 BC) opened up a new chapter of observation and pragmatism to which reference will be made later, but mention should be made of two contributions which are not only specifically relevant to our study, but which also set off a striking polarity of thought. Empedocles of Sicily (490–430 BC) postulated that the influence of the maternal imagination upon the embryo was so powerful that the development of the embryo could be guided and even interfered with; a statement made in a passage on the origin of twins and monsters.

The idea may recall a superstition held among Austrian peasant women that when a mother-to-be is frightened by a hare, her child will be born with a hare-lip. In reality, the mother is only frightened by the appearance of a hare, when the child she is carrying has already formed a hare-lip. Not only does the maternal imagination work upon the embryo, but elements in the embryo's development work upon the mother's imagination.

Democritos of Thrace (c.460–c.377 BC) held that it is the maternal womb that gives the human shape to the foetus. This idea received the attention of Aristotle, as well as being an integral element in contemporary concepts in the field of embryology and will be discussed in later chapters.

Stress on the importance of the mother's imagination as well as of the significance of the womb in giving the human form to the embryo is in stark contrast to views generally held at the time. For instance, Aeschylus (525–456 BC) in his drama *The Eumenides* has Apollo defend Orestes against the charge of matricide with the words: 'The mother of what is called her child is no parent of it, but nurse only of the young life that is sown in her. The parent is male and she but a stranger, a friend, who if fate spares his plant, preserves it till it puts forth.'

This would seem to be a very patriarchal denial of the woman's role in procreation, but also suggests the archetype of Gaia, the great Earth Mother in whom all seed is sown to germinate, grow and develop. Even so, all patriarchal societies relate procreation predominantly if not exclusively to the male; the Egyptians whose earth was male and heaven female and whose hymn to the sun begins with the words: 'Creator of the germ in woman . . .' later regarded the father alone to be the author of generation, the mother merely providing nidus and nourishment for the foetus.

In the Old and New Testament, only the male genealogy of Jesus or others is considered; female ancestry is rarely referred to. In this context, it is interesting to note that the Melanesian Trobriand islanders held that children originate in the spirit-world, Tuma, and incarnate into the menstrual blood where it has accumulated in the woman's body. They do not acknowledge the fertilizing power of the male semen nor the relation of sexual intercourse to conception and birth.

In returning to the Hippocratic school of medicine, we must distinguish between it and subsequent Hippocratic writings by pupils, but in all these we find interesting statements on embryology based on careful and astonishingly practical obervations.

The difference between male and female seed is described, the latter being regarded as a vaginal secretion. The substance of the embryo being humid, to begin with, is shaped by fire and draws nourishment from the food and breath taken in by the mother. Gradually, the fire dries up the original porous humid body and a more solid outer crust is formed. Then the heat begins to consume the interior humidity, giving rise to bones and nerves. From the remaining soft, wet parts, the inner channels are formed, the most humid of these giving the large blood vessels. In the intermediate parts, the remaining water contracts, giving rise to the flesh. The umbilical cord was recognized as the organ of foetal respiration.

In Hippocratic writings, it is stated that everything in the embryo is formed simultaneously; all limbs and organs take shape at the same time — those that are larger appearing more obviously than the smaller. An interesting theory was developed that explains the timing of birth by the exhaustion of the food supply to the growing foetus. Towards the end of one particular treatise it says:

> In a word, all the constitution of the foetus as I have
> described it to you, you will find from one end to the other
> if you will use the following proof: Take twenty eggs or more
> and give them to two or three hens to incubate. Then each
> day from the second onwards till the time of hatching take
> out an egg, break it and examine it. You will find everything
> as I say; in so far as a bird can resemble a man.

The mighty works of Aristotle (384–322 BC) which laid the foundation of the philosophical and scientific developments of our European culture, contain a wealth of new concepts in embryology, postulated in particular in his *On the Generation of Animals*. The first book of this work includes a discussion on the male and female roles in embryonic development. On the basis of the doctrine of Form and Matter and denying any seed-quality in the female share, it describes the catamenia, the menstrual blood, as the passive, plastic substance to which the dynamic, creative male semen gives shape and structure as the sculptor shapes the marble. The working of the male semen on the female blood in the uterus is likened here, too, to the action of rennet upon milk, solidifying the humid elements into drier, solid shapes.

The second book contains a vital discussion on the central principle of embryogenesis: How does a highly complicated and differentiated organism arise out of relatively simple undifferentiated elements? Is the ultimate form already present in the original matter (preformation), or does the ultimate form evolve step by step (epigenesis)? Aristotle was the first to put this question. He asks (Needham 1934, 28f):

How then does it [the shaping influence] make the other parts? All the parts, as heart, lung, liver, eye, and all the rest, come into being either together or in succession, as it is said in the verse ascribed to Orpheus, for there he says that an animal comes into being in the same way as the knitting of a net. That the former is not the fact is plain even to the senses, for some of the parts are clearly visible as already existing in the embryo while others are not; that it is not because their being too small that they are not visible is clear; for the lung is of greater size than the heart, and yet appears later than the heart in the original development.

This passage demonstrates the fact that Aristotle must have more minutely observed the step-by-step early development of hens' eggs than his predecessors, and in so doing corrected a Hippocratic theory.

After suggesting that the catamenia be likened to an automatic machine or clockwork set into motion by the dynamic inherent in the semen, Aristotle goes on (Needham, 1934, 30):

But how is each part formed? We must answer this by starting in the first instance from the principle that, in all products of Nature or art, a thing is made by something actually existing out of that which is potentially the same as the finished product. Now the semen is of such nature and has in it such a principle of motion, that when the motion is ceasing each of the parts comes into being and that as a part having life or soul . . . And just as we should not say that an axe or other instrument or organ was made by the fire alone, so neither shall we say that foot or hand was made by fire alone . . . While, then, we may allow that hardness and softness, . . . and whatever other qualities are found in the parts that have life and soul, may be caused by mere heat or cold yet, when we come to the principle [*logos*] in virtue of which flesh is flesh and bone is bone, that is no longer so; what makes them is the movement set up by the male parent, who is in actuality what that out of which the offspring is made is in potentiality. That is what we find in the products of art. . . . For the art is the starting-point and form of the product; only it exists in something else

(i.e. potentially in the mind of the artist), whereas the
movement of nature exists in the product itself, issuing
from another Nature (i.e. the parent) which has the form in
actuality.

Within his formidable concept of the duality of form and matter,
Aristotle not only formulates the two basic ideas of embryogenesis
— preformation and epigenesis — but also, although quite clearly
holding that the principle form element (Logos) is epigenetic in
nature, he seems to concede a degree of preformation to primary
qualities. Just this we shall recognize later on as the essence of
modern thinking in the field of embryology.

Aristotle develops the idea of a step-by-step ensoulment of the
embryo. Already catamenia and semen are endowed with a vegetative soul: 'First of all such embryos seem to live the life of a plant.
. . . As it develops, it also acquires the sensitive soul in virtue
of which an animal is an animal.' As a last step, the rational soul
peculiar to man connects itself with the foetus. 'All three kinds
of soul must be possessed potentially before they are possessed
actually. . . . The end is developed last, and the peculiar character
of the species is the end of the generation in each individual.'

The idea of successive steps in the ensoulment of the embryo
has dominated theological embryology throughout its history and
was made the basis of legal arguments concerning abortion. It also
seems to contain elements that led to the theory of recapitulation
propounded by the German thinker, Haeckel, in the nineteenth
century, which held that the human embryo repeats in its development phases of phylogenetic evolution.

It is not only in biology that we encounter the relevance of
Aristotle's thinking for our own time. It is equally the case, for
instance, in physics, especially in relation to the twofold nature
of light as wave as well as of corpuscle, which still presents an
intellectual riddle. The negligible time-span between Plato and
Aristotle marks the division between two ages: the age of
meaning, of mythos, from the age of cause and necessity in
scientific thinking. Yet Aristotle still endeavoured to include in
his commitment to causal thinking an element of meaning and
significance as the 'final cause'. He criticizes the natural philosophers of his time by saying (Needham 1934, 38):

> [They] did not see that the causes were numerous; they only
> saw the material and efficient causes and did not distinguish
> even these; while they made no inquiry at all into the formal
> and final causes . . . Democritos, neglecting the final cause,
> reduces to necessity all the operations of Nature. Now they
> are necessary, it is true, but yet they are also for a final
> cause, and for the sake of what is best in each case.

Few people today would be moved by Aristotle's inclusion of
the question of what is best and good in the scientific study of
nature. To most of us this would seem to be a lost cause. The
course of the last two thousand years has led to a purely objective
science of necessity, which holds that it must have no part in the
quest for the good.

The Stoic and Epicurean schools of philosophy of the third and
second centuries BC seem to have adopted the earlier views of
Hippocrates and held that the male and female shares in embryogenesis were equal and that all parts of the embryo were formed
out simultaneously. In connection with Epicurean Atomism, there
was a tendency towards a preformationist view. However, only
in the first Christian century do we find in Seneca's *Naturales
Quaestiones* a clear and definitive statement on preformation: 'In
the seed are enclosed all the parts of the body of the man that
shall be formed. The infant that is borne in his mother's womb
has the roots of the beard and the hair that he shall wear one
day. In this little mass likewise are all the lineaments of the body
and all that which posterity shall discover in him.' It is interesting
that this very first pronouncement on Preformation is so categoric
and extreme. We shall see the same trend recurring increasingly
in the centuries that follow the Renaissance.

In the second century Galen of Pergamos (*c*.AD 130–*c*.200) gathered up most of the fruits of the medical knowledge of antiquity,
but in his treatise on the *Natural Faculties* and on the *Formation
of the Foetus*, he developed an original theory of embryogenesis
(Needham 1934, 52f):

> Genesis is not a simple activity of Nature, but is compounded
> of *alteration* and of *shaping*. That is to say, in order that

bone, nerve, veins, and all other tissues may come into existence, the underlying substances from which the animal springs must be altered; and in order that the substance so altered may acquire its appropriate shape and position, its cavities, outgrowths, and attachments, and so forth, it has to undergo a shaping or formative process.

Going into more detail, Galen continues:

The seed having been cast into the womb or into the earth — for there is no difference — then after a certain definite period a great number of parts become constituted in the substance which is being generated; these differ as regards moisture, dryness, coldness and warmth, and in all other qualities which naturally derive therefrom . . . Thus the special alterative faculties in each animal are of the same number as the elementary parts, and further, the activities must necessarily correspond each to one of the special parts . . . while the bringing of these together, the combination therewith of the structures that are inserted into them, etc. have all been determined by a faculty which we call the shaping or formative faculty; this faculty we also state to be artistic — nay, the best and highest art. . . .

While Galen's opening sentences concerning the identity of womb and earth refer back to archaic images, his distinction between the specialized parts and the unifying form is akin to contemporary concepts.

Another original idea of Galen is that of the *Tria Principia* which states that during the early stage of embryonic development three organs initially appear out of the unformed seminal stage (this was later described by Ambroise Paré (1510–1590) in the Renaissance as three bubbles or bladders) — the heart, the liver and the brain. One may wonder whether the three germ layers of ectoderm (brain), mesoderm (heart) and endoderm (liver) were intuitively divined here.

With the Hippocratic writings as basis, Galen gives a wealth of often astonishingly accurate descriptions of embryonic features as well as of the structures of human and animal bodies. In his theological outlook, he stresses the perfection of adaptation and the inner necessity of all organic manifestations. At the conclusion of his massive work, he says: 'Such then and so great being the value of the argument now completed, this makes it all plain and clear like a good epode . . . a hymn to the Gods.'

The Church fathers of the early Christian centuries, like Tertullian, St Augustine and St Gregory for example, seem to have been influenced by pre-Aristotelian views in their concern with the question of the ensoulment of the foetus. Tertullian (c.160–c.222) maintained that the soul was fully present in the embryo from the start; St Augustine (354–430) suggested that ensoulment occurred between the second and fourth months. For a time, a distinction was made between the *embryo informatus* which had no soul, and the *embryo formatus* which had, the demarcation being the fortieth day for males, the eightieth for females. (It is now established that by the sixtieth day, the human *form* is clearly apparent in the embryo.)

Gradually, Roman law which like Spartan law allowed abortion (theoretically recommended by Plato and incorporated into the law in Athens by Lysias) was replaced by a decided condemnation of all pre-natal infanticide.

The thinking of men in these early Christian centuries was still under the influence of the neo-Platonist school, and Origens' doctrine of pre-existence was still widely acknowledged. This latter idea held that all human souls were created at the beginning before the creation of matter, souls incarnating into the seed-blood mixtures in rotation as these were being prepared. This theory, however, was opposed by the idea of creationism, largely held by the Eastern Church, which stated that while the body derived from Adam by propagation, the soul was each time newly created by God and implanted into the embryo.

A third school of thought, often referred to as traducianism, and mainly upheld by the Western Church fathers, taught that both souls and bodies came from Adam and were passed on from parent to child. Tertullian as well as St Augustine supported this latter doctrine.

St Gregory of Nyssa (c.332–c.396) seems to link the concept of preformation to the manifestation of the soul, stating that the soul makes its body as if it were a gem making a stamp upon some soft substance and acting during development from within (Needham 1934, 58f): 'for we do not suppose it possible . . . that a certain seal should agree with a different impression made in wax'. 'The thing so implanted by the male in the female is fashioned into the different varieties of limbs and interior organs, not by importation of any other power from without, but by the power which resides in it transforming it.' He apparently held the soul to be inherent in the seed.

Talmudic writings of the second to the sixth centuries contain various references to foetal development. The writers were obviously conversant with Aristotle's and Galen's works, adopting the distinction between those parts that stem from the father's semen and those that come from the mother's blood, the father who sows the white elements, the mother who sows the red. But two further elements are added, both of considerable antiquity: the duality of the male and female share in propagation is extended to include a trinity – God who contributes the life, the soul and the expression of the face to the thing in becoming, God who hovers above and between male and female.

In the encounter with neo-Platonism, the cabbala revives the age-old image of Adam Kadmon, the androgynous macrocosmic man in the idea of Adam Protoplastes, linking man as a microcosm into the universe. Later on in the seventeenth century, these writings were to inspire some extraordinary attempts in the field of speculative embryology on the part of personalities such as Leibnitz, van Helmont, Marci and Borelli.

In the thirteen century, the two great thinkers, Albertus Magnus and Thomas Aquinas, made Aristotle's work the foundation of scholasticism. Albertus (d. 1280), a highly learned and knowledgeable man, not only revived Aristotle's ideas on embryology but made considerable contributions of his own, especially to animal embryology. He had observed the pulsating blood-islands in three-day-old chick embryos and related them to the development of the heart. Contrary to Aristotle, he held the view of earlier Greek philosophers and the Epicurean School that seed was not only bestowed by the male but also by the female. Yet he also stated that to these two humidities a third, that of the menstrual blood, had to be added. He thought that the female seed underwent coagulation like cheese through the impact of the male seed. He did, however, share Aristotle's views on epigenesis and preformation and saw in the male semen the actual shaping power working upon the menstrual blood as its material. With Albertus Magnus, a new spirit of inquiry dawned, having died down for the millenium that followed Galen of Pergamos, to flare up and become a burning torch two hundred years later in Leonardo da Vinci.

Thomas Aquinas (1225–1274) embodied Aristotle's philosophy in his *Summa Theologica* and held 'the female generative power to be imperfect compared to the male just as in the crafts the inferior workman prepares the material and the more skilled one shapes it'. Also Aquinas' theory of foetal animation closely followed Aristotle's thinking. He postulated that the embryo was originally endowed with a vegetative soul which perished, giving way to a sensitive soul. This in turn died, its place being taken by a rational soul sent directly from God. This doctrine is beautifully expressed in a Persian Sufi poem:

> I died from mineral and plant became;
> Died from the plant and took a sentient frame;
> Died from the beast and donned a human dress;
> When by my dying did I e'er grow less?

In the Renaissance, Leonardo da Vinci (1452–1519) stands out as a thinker as Aristotle did in antiquity. He reduced Aristotle's principle of a fourfold causation in natural phenomena to the oneness of necessity. While many of the artists of his time carried out dissections of corpses in connection with their studies of the human body, only he developed the inquiry into the how and why of the subject he so acutely observed. Leonardo's embryological drawings and writings are in his third notebook among sketches of cogs and pulleys and beautiful anatomical drawings. They display extraordinary common sense and contain several original, apt and detailed descriptions of embryos. The weight-taking quality of the amniotic water, the function of the umbilical cord are recognized, and Leonardo introduces measurements of growth in a generally quantitive and mechanistic approach. While Aristotle deplored the monopoly of the principle of necessity, Leonardo embraced it with enthusiasm as the only natural law which governs the world: 'Necessity is nature's master and guardian; it is Necessity that makes eternal laws.' (Needham 1934, 81).

The work of three Italian scientists born in three successive years — Aldrovandus (1522–c.1605). Fallopius (1523–1562) and Eustachius (1524–1574) — is mentioned by Karl König (1967) as of special importance to the development of modern embryology.

After Aristotle, Aldrovandus was the first to examine incubating hen's eggs systematically and to describe their stage-by-stage development. He discusses in full and also criticises Aristotle's earlier writings as well as those of Galen and Albertus Magnus and says that in order to bring to an end the controversy between philosophers and physicians concerning preformation and epigenesis, he pursued with keenest curiosity and diligence the incubation of twenty-two hen's eggs, opening one each day. He found Aristotle's doctrine to be the most likely as a result.

Aldrovandus' writings which are voluble and literary, are replete with new and valuable observations and scientific arguments. Fallopius and Eustachius, on the other hand, are better known for their anatomical work. Fallopius discovered the tubes leading from the uterus to the ovaries, which henceforth have borne his name, and Eustachius the corresponding organ in the throat which are the tubes leading from the pharynx to the middle ears, called Eustachian tubes. We shall later on encounter a particular significance in these two organs for the mythos of embryogenesis.

The seventeenth century can be seen as a seething cauldron containing the most diverse and singular ideas. In 1625, Aromatari (c.1586–1660), likewise an Italian, in a publication on plant development, postulated the principle of preformation for the first time since Seneca. Aromatari had observed that in plant bulbs and seeds, rudiments of the adult plant could be discerned, and he concluded that probably in all plants as well as in animals the embryo form would be roughly sketched out in the seed or egg.

Other authors of the time re-established the fundamental morphological facts of embryology, holding various views, some vitalistic, ascribing the primary cause of development to God, to the *anima mundi*, or to the individual soul: others taking a more naturalistic view like K. Digby (1603–65), or an atomistic orientation like Highmore (1613–85).

The most central contribution to embryology in the seventeenth century came from Harvey (1578–1657) in his *Generation of Animals*. Harvey was an outspoken vitalist, holding that not only differentiation in form but also in growth were expressions of an immanent spirit. He adhered to most of Aristotle's teachings, particularly to the idea of epigenesis, but took it further by adding to the picture of the stone-mason shaping pre-existent material, that of the sculptor who builds up a form by adding clay to it. He distinguished 'differentiation *without* growth', corresponding to Aristotle's theory that semen fashions the catamenia in the womb, from 'differentiation *with* growth', (epigenesis) which he held to be applicable to all sanguineous animals.

Harvey's most important contribution to embryology, however, was his statement that all living organisms developed from eggs, which led to the formulation *omne vivum ex ovo*. This hypothesis was partly based on his profound scepticism of the idea of spontaneous generation, and partly on the results of his observations of small embryos surrounded by their chorion which he mistook for eggs (the ovarian follicle was not discovered until fifteen years after his death and the actual ovum not before the nineteenth century).

It was the development of the microscope that led to the discovery of the ovarian follicle by de Graaf (1641–73) in 1672, which at the time was taken for the ovum, and in 1677 to the discovery of the sperm by Leeuwenhoek (1632–1723). Both these discoveries were to give rise to most bizarre theories in the century that followed.

Novel and interesting attempts were made by Gassendi (1592–1655) and by Descartes (1596–1650) to explain embryology as mechanical processes of movements and pressures in a mathematical-geometric theory against the background of an atomistic, iatromathematical outlook. While these obviously very premature attempts did not lead any further, they can be seen as forerunners of contemporary approaches to the subject of embryology like those of Blechschmidt to which references will be made in Chapter 6.

A very different set of ideas, however, can be discerned in connection with atomistic embryology. In van Helmont's and Leibniz's writings, elements of neo-Platonism and the cabbala appear. In the former's *Doctrine of Seeds*, the idea that a portion of matter endowed with a spiritual *Archaeus*, the *aura vitalis*, re-emerges. It is based on the neo-Platonist emanation theory of a world of ideas and beings (in Greek *daemonions*) which were the spiritual representation of all earthly things and beings, and the ancient concept of the correspondence between microcosm and macrocosm, of Adam Protoplastes in whom all men are essentially contained.

The cabbala, which emphasizes the difference between individual persons rather than their likenesses, builds a link to the idea of preformation that begins to reappear in the next century, but the seeds, the *aura vitalis* alone, corresponding to Aristotle's *anima vegetiva*, cannot create a complete being. The *lumen formale et vitale*, the element of light, issuing directly from God, is required.

The idea of light as the forming element in embryology was taken up in a fascinating way by M. Marci of Kronland (1595–1667) in a thesis, *Idearum operatricium Idea*, in 1634. Partly a purely scientific contribution to optics, it also constitutes a highly speculative theory of embryogenesis. It explains the manifold and complex development of form from seed to organism by means of the analogy of lenses which produce complicated rays and patterns of light from a simple light source. The formative power of light is imagined as radiating from the geometrical centre of the foetal body, creating the differentiated complexity of form without losing any of its own power. In addition to what has just been mentioned, Marci's thesis may well be linked to the awakening interest in perspective in the previous centuries in Italy and Germany as well as to the emergence of a new perspective geometry in the following century. We shall encounter this connection again in Chapter 6 when we hear of the contemporary work of Lawrence Edwards on form-development.

Of similar relevance to the most modern thinking in the field of embryology, and equally void of any response in its own time, is *De Motu Animalium* by Borelli (1608–79) in which the function of the semen is compared to a magnet arranging iron particles in a field of force. Before Borelli, Harvey in his paper on conception, published in 1653, states: 'The woman does seem, after spermatical contact, to be affected in the same manner, and to be rendered prolifical, as the iron touched by the loadstone and presently endowed with the virtue of the loadstone.' Here again the connection to modern concepts of morphogenetic and individuation fields is apparent.

In 1674, John Mayow (1640–79) published a treatise on the respiration of the foetus in which in a brilliant and elegant argument based on all the then available evidence, he reinstated Hippocrates' long discarded idea of foetal respiration through the umbilical cord. His ideas were based on the concept of the resorption of gases; in modern terms, and together with Robert Boyle's contributions to chemistry, his work stands out in clarity from the often confused thinking of the seventeenth century.

By far the most intense and immediate influence on ideas of embryology, however, came from Marcello Malpighi's (1628–94) treatise, *De Ovo Incubato* published in 1672, describing and beautifully illustrating his microscopic examinations of newly fertilized hens' eggs, invariably displaying differentiated and discernibly embryonic forms. This led Malpighi to adopt a definitive preformationist stand. While he acknowledged non-simultaneous unfolding and varying growth rates in the embryo, he held that form and differentiation are already present in the fertilized ovum, only to unfold in the process of incubation. His observations which, as he reports, were carried out during a hot August in Italy were, of course, of laid eggs which had after fertilization passed through the warm oviduct and had, therefore, already gone through the first phase of development, so that among the laid eggs, he could never find one without the first beginnings of embryonic development. While Malpighi's descriptions and illustrations of early chick development were of fundamental importance, his moderately and reasonably held view of preformation had the most bizarre consequences.

2. A BRIEF HISTORY OF EMBRYOLOGY

At that time, Swammerdam (1637–80) in Holland, having observed a fully developed butterfly folded up in a chrysalis, concluded that the butterfly was already hidden (*larvatus*) in the caterpillar and therefore also in the egg. He expanded his theory by saying: 'In Nature there is no generation, only propagation, the growth of parts. All men were contained in the organs of Adam and Eve; so original sin is explained. When their store of eggs is exhausted, the human race will cease to exist.' On the other hand, Swammerdam made contributions of lasting importance to biology, particularly with his work on amphibia and especially with his description of the cleavage of the egg cell and its ensuing segmentation.

By 1684, a scientist was reported to have seen minute embryos in unfertilized eggs and at the turn of the seventeenth century, members of the public as well as the clergy began to take up the idea of preformation with all its philosophical and religious implications. It seemed to satisfy some peculiar need or expectation in human minds at the time.

Maitrejean, the French surgeon of the seventeenth century, had his admirably illustrated work on the development of the chick published posthumously in 1722. In this work he attributed Malpighi's observations of normal early incubation and development to the heat of an Italian summer, thereby dismantling Malpighi's preformation argument and outspokenly upholding the principle of epigenesis. In spite of this, the idea of preformation spread like wildfire throughout that century. This spread was not only fanned by the work of Malpighi and Swammerdam, but also by publications on the part of men like Andry, Delenpatius and Gautier who, following on Leeuwenhoek's earlier discoveries, claimed to have seen under the microscope minute humans inside the spermatozoa which they called 'animalcules' complete with heads, arms and legs. It was even maintained that a horse was seen in horses' semen, a long-eared donkey in donkey's semen and a cock in cock's semen! There are drawings showing little humans, either with outstretched arms and legs or crouching with updrawn knees, embraced by the arms like embryos.

Just as Leeuwenhoek's discovery of the spermatozoa gave rise to 'animalculism' so did de Graaf's discovery of the ovarian follicle, which was taken to be the ovum, give rise to 'ovism'. While no one claimed to have seen human embryo forms in the ovum, the theory of ovist preformation had notable adherents like Malpighi, Swammerdam, Bonnet, Haller, Vallisneri, Spallanzaini and many others. They all held that the embryo was preformed in the ovum and that Eve's body contained all the embryos that were ever going to be born. Animalculism was supported by a smaller group of men, but among them were thinkers like Leeuwenhoek, Leibniz and Cardinal Polignac.

While the controversy between ovists and animalculists engaged a great number of minds and occupied a prominent place in the thinking of the eighteenth century, the much more essential question was that of preformation versus epigenesis. In all the disputes that resulted, de Buffon (1707–88) appears as a rather independent figure, but his ideas are difficult to assess and seem at times odd and highly theoretical. He upheld against both ovists as well as animalculists the idea first expounded by Empedocles that while the foetus may originate from ovum or semen, it was moulded into its human form by the maternal organism during pregnancy.

One of the few outspoken protagonists of epigenesis after Maitrejean was J. T. Needham (1713–81) whose interest in the spontaneous generation of microscopic organisms and whose adoption of the idea of a vegetative vital force set out in Leibniz's *Monadologie*, may have contributed to the linking of epigenesis to vitalism. In his *Idée Sommaire* published in 1776, he writes:

> The numerous absurdities which exist in the opinion of pre-existent germs together with the impossibility of explaining on that ground the birth of monsters and hybrids, made me embrace the ancient system of epigenesis, which is that of Aristotle, Hippocrates and all the ancient philosophers, as well as of Bacon and others.

Maupertuis (1698–1759) also supported the epigenetic approach when he likened the appearance of form in the foetus to the *Arbor Dianae*, the plantlike coagulation that comes about in the mixture of silver and spirits of nitrate with mercury and water.

The most important representative of the theory of epigenesis in the eighteenth century was K. F. Wolff (1733–1794) who in his *Theoria generationis*, published in 1759 when he was only twenty-six years of age, presented a philosophical argument in its favour, adopting Leibniz's idea of the monad, but investing it with an inherent force, the *vis essentialis*. He argued that if an organism pre-exists, if all the organs were already present and only invisible because of their minuteness, they would have to have their complete and finished form as soon as they became visible. If, on the other hand, organs are not preformed, then one ought to be able to witness one shape changing into another. He watched the development of the tip of a growing plant through the microscope, making careful drawings of what he observed. Shoot after shoot displayed only homogenous tissue. There was no sign in this tissue of the leaves, blossoms and other organs which were later to emerge from the particular plant. He noticed, however, that when the specialized parts did begin to form, they first appeared as a simple prominence or swelling in the undifferentiated tissues.

When Wolff subsequently published similar observations of the gradual appearance of new organs and the development of blood vessels in the chick embryo, he called forth intense opposition on the part of Albrecht von Haller (1708–77), professor at Göttingen in Germany. Von Haller who had originally held epigenetic views, had become the most explicit protagonist of preformation which, together with ovism, had become the generally accepted outlook at the time.

Von Haller's lasting contributions to embryology were his massive publications critically examining much of the work done in the preceding fifty years as well as his original work — the introduction of exact measurements and comparisons of embryonic growth rates. His argument was basically that the chick embryo was so fluid in its early stages that Wolff had no right to deny the presence of a given structure, simply because he could not see it. Wolff responded with a description of the formation of the yolk membranes and their gradual vascularization, to which von Haller answered: 'I don't believe that any new vessels arise at all, but that the blood which enters them makes them more obvious because of the colour it gives them.' Wolff's reply was the publication of *De Formatione Intestorium* in 1768 in which he demonstrated that the chick intestine is formed by the folding back of a sheet of tissue which is detached from the ventral surface of the embryo, and that the folds produce a gutter which transforms itself into a closed tube. The intestine, therefore, could not possibly be preformed, as von Haller had stated, and Wolff proposed to ascribe the same process to other organs, while developing on the basis of these observations a general theory of epigenetic development for all organs. There is no doubt that this theory is equally applicable to the formation of the neural tube, which marks the beginning of the development of the nervous system.

Von Haller's objections to epigenesis were not limited to the argument that the non-appearance of a structure does not exclude its 'effective' or 'potential' presence. He also made a very cogent objection to epigenesis by asking how Wolff's *vis essentialis*, which was only one force, made the material issuing from a hen always produce a chicken, that from a peacock a peacock and so on, thus pointing to the problem of heredity which, first discovered by Mendel in 1866, was only understood in our own century. While Wolff's observations have never been disproved and really undermined the preformation theory, the impact of his work was not felt, as von Haller had by far the greater name and influence in his time, so that preformationism and even ovism held the biological scene.

Although Wolff's work had so limited an influence, to begin with, it did lead J. F. Blumenbach (1752–1840), who came from a preformationist outlook, to a deep appreciation of Wolff's ideas, which he expressed in his thesis *Über den Bildungstrieb* published in 1789. He took up Wolff's *vis essentialis* as his *nisus formativus*, which he imagined as an active, constant force, giving shape and form to living bodies. Kant in his *Critique of Judgment* mentions both Wolff and Blumenbach and himself adopted the epigenetic theory of embryogeny.

Before the end of the eighteenth century, Goethe (1749–1832) had developed his unique and original ideas on morphology and metamorphosis. The German philosopher, Herder, a close friend, had developed the ideas of both Aristotle and Thomas Aquinas on the consecutive appearance of vegetative, sentient and rational souls in embryogenesis into a hypothesis, proclaiming one evolution for all organisms from plants to animals to man, an idea that ultimately became the cornerstone of modern biology through Darwin and his followers.

While Goethe was preoccupied with this theory, he also studied Linnaeus' system of botany with keen interest 'striving out of the innermost need of my being to unite what Linnaeus had separated'. Goethe's own approach was exceptional. As a poet and artist, he experienced that truth resulted when sense-perception and idea fused in cognition:

> When man's healthy natural faculties co-operate as a totality;
> when man feels the world about him as a great, beautiful,
> dignified and spacious whole; when this sense of ease and
> harmony culminates in a state of pure, free delight — then
> the universe, if it were endowed with feeling, would exult
> in a sense of final achievement and admire the crowning
> glory of its own evolving essence. (*Wisdom and Experience*).

We can see how fundamentally different, in fact diametrically opposed, is Goethe's method of gaining understanding from the modern reductionist trends in objective science. We shall return to this issue later on. Goethe sought the *idea* in Nature where it was clearly revealed and then descended from the more developed to the simple. He did not try to explain the composite from the simple, but rather endeavoured to perceive the complete or whole at a glance and to understand the less fully-developed as a one-sided derivation from the complete or perfect. Nor did he seek for extraneous, mechanical causes that form the plant, but for the innate idea and necessity that leads to the development of form-differentiation in a particular organism.

The *archetypal plant* was to Goethe not only a concept, but an idea he could intuitively perceive in each and every plant. He describes the archetypal plant as the leaf that expands from the node and contracts again into its point. The principle of contraction manifests in seed and bud, rhythmically interacting with that of expansion in leaf and stem.

Goethe maintained that the archetypal plant not only enabled him to recognize every plant as a specific expression of this principle, but also to imagine non-existent plants that, given the required conditions, could exist. Thus he was able to explain the propagation of the plant, Bryophyllum, which generates new plants at the tips of its leaves — the points of contraction — equivalent to the seed. He subsequently predicted a phenomenon that he discovered later — namely, a secondary rose blossom growing out of another blossom.

For the rest of his long life, Goethe tried to find a corresponding archetype for the animal kingdom. In his *Outline to a Comparative Anatomy* published in 1796, he says: 'This then have we gained: we can boldly state that all higher organisms such as fishes, amphibia, birds, mammals and, at their summit, man, are formed according to one archetype which only in its permanent parts inclines more to one or another side. . . .'

In his search for the vertebrate archetype as the link between man's head-brain predominance and the animal's body predominance, Goethe was led to see the central position of the vertebrae. This in turn led him to the discovery of man's intermaxillary bone, the existence of which his contemporaries still denied, and he recognized several of the other skull bones as metamorphosed vertebrae.

Goethe's contributions to an understanding of natural phenomena, however, did not lie in his factual discoveries, but rather in the completely new method of approach he established as 'intuitive discernment' or 'contemplative judgment' which allowed the wholeness of organisms to be perceived and experienced. This method, while largely disregarded or even rejected throughout the nineteenth century, was taken up and developed by Rudolf Steiner at the turn of the century and has emerged in our own century, not only in the work of many of Rudolf Steiner's pupils, but also in P. Weiss's ideas on the hierarchical structure of organisms, Waddington's conception of the competence of tissues, or Koestler's idea of holons. All these possess a certain quality of the archetype and we shall return to them later and shall finally take up Goethe's idea of the vertebrate archetype as mythos.

Current Thinking
in Biology and
Embryology

3. Mendel and Darwin

Modern man's conception of himself is still rooted in the theories of Mendel and Darwin of the nineteenth century. It seems, therefore, appropriate to include the thinking of that century in contemplating contemporary ideas in the field of biology and embryology, although it was still influenced by the conflict between the idea of a divine creation, upheld by the churches, and the then new ideas on evolution held by science.

Lamarck's *Philosophie Zoologique* published in 1809 opens up the theme of evolution with new theories of growth, differentiation, development and generation. Lamarck (1744–1829) postulated that nature tends to increase the size of living organisms to a pre-determined limit; that the creation of new organisms is brought about by new necessities; that the degree of development in a new organ is proportional to the degree to which it is useful; and finally, that everything *acquired* by an individual is transmitted to its offspring. Charles Darwin's grandfather, Erasmus Darwin, had held similar views before Lamarck, and considered that environment in general was responsible for inherited changes.

It was, however, Charles Darwin's *Origin of Species* published in 1859 as well as his *Descent of Man* that took hold of the thinking of the time. Darwin's (1809–82) work was based on meticulous observation and brilliant reasoning, his material having been gathered on his five years of exploration in South America and its islands, on the *Beagle* from 1831 to 1836. His evidence was partly derived from geological variations in the flora and fauna of the islands and the South American continent.

After Darwin's return to England, he developed his theory of natural selection under the influence of Malthus's *Essay on Population*, which predicted a worldwide struggle for survival as the human population would increase in geometrical progression and ultimately outgrow the earth's potential food supply, hence, the original title of Darwin's famous work was *The Origin of Species by Means of Natural Selection or the Preservation of Favoured Races in the Struggle for Life*.

Darwin was aware that his theory of evolution by natural selection depended on a sufficient potential of variability in the vast abundance of organisms. He also realized that the blending of characteristics which was the generally accepted view of his time, would inevitably imply a continuous reduction in variation and an increasing uniformity of species.

Mendel's (1822–84) incisive discovery of the basic laws of heredity, published in 1866, went unnoticed for the remainder of the nineteenth century. In many years of careful plant-breeding experiments, Mendel established that heredity is transmitted by a considerable number of independent factors. When each parent contributes the same factor, a constant character is produced in the offspring. When each parent furnishes a different factor, a hybrid results which bears the features (the phenotype) of the dominant factor. When the hybrid procreates in turn, the two different factors are liberated, producing phenotype progeny in a proportion of 3:1 of dominant and recessive factors respectively. Thus, while hybrids with 1 pair of factors produce 3 genotypes (factor combinations), hybrids with 2 pairs of factors produce 9, with 3 pairs 27 and so on.

As living organisms usually carry very many hereditary factors (now called genes), a degree of hybridization is inevitable and the potential of variation, though appearing gradually, is unlimited.

Mendel's discoveries, which were not only to provide the basis for Darwin's theory of evolution, but also for the development of the new science of genetics, lay dormant, to be rediscovered only at the beginning of the present century by de Vries in Holland, Correns in Germany, and Tschermak in Austria.

In 1874, Haeckel (1834–1919), the great German biologist and an enthusiastic adherent of Darwin's theories, introduced the idea of evolution into embryology and took up the earlier theories of recapitulation from the work of Meckel, Sévres and von Baer. He formulated the 'fundamental biogenetic law': that ontogeny, which is the embryonic development of the individual, was a recapitulation of phylogeny, which is the evolution of the species as a whole.

This idea together with the cliché, 'the phylogenetic tree', seemed to provide a world conception at the time which probably met some deep need of meaning and significance in human existence.

Like K. F. Wolff, Haeckel seemed to possess some of the vision of Goethe's archetypes, of the idea as visible form. While today, ontogeny cannot be regarded as the actual recapitulation of phylogeny, Haeckel's efforts in the field of biology still inspire contemporary interpretations of embryogenesis.

In our own century, Karl König (1927, 1928 and 1967) Kaspar Appenzeller (1976) and Frits Wilmar (1979) developed the idea of human embryogeny as images of earth evolution in connection with Rudolf Steiner's teaching, Anthroposophy. In the third part of this book, these ideas will be taken up.

4. The beginnings of experimental embryology

Thinking in the field of biology and embryology at the end of the nineteenth and beginning of the twentieth centuries was again subject to the impact of the controversy over epigenesis and preformation. Decisive contributions here came from a new branch of science, experimental embryology.

The renewed controversy, however, was opened by the speculations of a German zoologist, A. Weismann, with his germ-plasm theory. He held that the separate parts into which the nucleus of the ovum breaks up in cell-division are separate from each other like the bits of a mosaic and develop independently, albeit in perfect harmony with each other, into the finished organism. Weismann called the separate bits 'primordia' (later named chromosomes), and regarded them as the starting-points of organs and tissues. This ingenious hypothesis, based purely on intuition, preceded the discovery and grasp of the chromosomes by half a century.

Wilhelm Roux, anatomist at the University of Breslau, decided to test Weismann's theories. Were they to prove correct, an embryo at the two-cell stage would contain one half of the individual in each cell. If one of the two cells at this stage could be removed while maintaining life in the other, the other cell would have to develop into half an individual.

Roux achieved his historic experiment in 1888. Taking a two-cell embryo of a frog, he destroyed one of the cells with a heated needle and managed to keep the other cell alive and to cultivate it further, in its watery element. The result was a mutilated half-tadpole. This experiment seemed to prove that each cell had contained one half of the means of construction of a frog, and the experiment was regarded as proof of Weismann's theory and with it, of preformation.

Yet there were those who doubted the validity of the proof, pointing out that the destroyed cell had remained attached to the surviving one and could possibly have influenced its development. Various attempts were made to separate embryonic cells, until in 1891, Hans Driesch of Germany, achieved it in the following way: Driesch spun two-cell sea urchin larvae in sea water in a centrifuge and succeeded in separating them without injury to either and to incubate each cell further into complete sea urchin larvae. Later on, he achieved the separation of four-cell embryos and obtained from each cell on incubation, a complete creature, albeit smaller than average. Driesch also demonstrated that one whole embryo could be obtained by the fusion of two eggs into one. The spectacular emergence of a whole embryo from its fragments was so contrary to the general preformationist outlook of the time that it triggered off an upsurge of experimental research at the beginning of the present century.

The most incisive contributions to modern embryological thinking were probably those of Hans Spemann in Germany. Following up the work of Roux and Driesch in testing Weismann's 'mosaic theory', Spemann tied a slip-noose of baby's hair across the fertilised egg of a newt, a kind of salamander. He observed that the half containing the nucleus developed, but if nuclear substance slipped through to the other half, the latter would also develop into a complete newt. Spemann noticed, however, that this happened only when the tie was made across a 'grey crescent', a darkish segment that appears on the surface of the newt's egg before cleavage. The piece which did not contain any part of the 'grey crescent' divided and sub-divided, but produced a disorganised shapeless mass of tissue which Spemann called 'belly piece', because it consisted only of liver, lung and intestinal cells as well as other abdominal material, without any axial skeleton, nervous system or unifying pattern.

The discovery of the part played by the grey crescent

subsequently gave rise to a new theory of pre-determination of embryonic structure, but Spemann himself was searching for a more lasting and essential feature, realizing that the grey crescent disappeared as soon as cells multiplied and was only to be observed in certain kinds of salamander.

In fact the normal development of amphibian eggs generally proceeds in the following way: the fertilized ovum divides into two, then four, then eight cells and so on, finally forming a cluster or morula of many cells. These then form a hollow, fluid-filled sphere which is called the blastula. The blastula begins to develop an indentation which becomes an invagination, proceeding further until a cup, the gastrula, is formed with an outer and an inner layer of cells which pass over into each other at a small opening, the blastopore. The blastopore is situated just below the area the grey crescent had occupied.

In 1924, Spemann demonstrated the organizing power of the blastopore in a series of fascinating experiments. He transplanted a piece of salamander gastrula from above the blastopore that would have become belly skin into the area below the blastopore of another salamander gastrula that would have become nerve tissue and vice versa. The cells that would have become belly skin became nerve tissue, and the cells which would have formed nerve became skin.

In another experiment, Spemann implanted the blastopore region of a colourless salamander gastrula onto a dark salamander larva. As the latter developed, the implanted blastopore called forth from the surrounding cells of the dark larva a second twin embryo, with a complete axial and nervous system.

Spemann called the blastopore the 'organizer'. He did not, however, suggest that it controlled the entire process of development, but rather envisaged a succession of sub-organizers as development proceeded. Soon organizers similar to the one he had discovered in salamander larvae were found in many other classes of vertebrate creatures. In all cases, the organizer principle seemed to be responsible not only for the main axis of the embryo, but also for the moulding of relatively late tissue and organ formations into an organic structure. The organizer appeared to be the epigenetic principle, but what *was* the organizer? How did it work?

It was natural to assume that functions of the living cells would generate either dynamic or metabolic processes which would induce differentiation. When in 1932, C. H. Waddington in Cambridge, and independently, J. Holtfreter in Germany, demonstrated in the chick and newt embryo respectively that blastopore tissue, even after having been effectively killed off, could still induce development, the way was open for the search for a chemical substance that acted as organizer. Several groups of researchers tried to identify the substance in repeated efforts to analyse blastopore tissue, but were frustrated by not finding one substance which would induce cell-differentiation and by the fact that so many substances did just that.

When two years later, Brachet, Needham and Waddington proved conclusively that methylene blue, a chemical dye which certainly cannot be considered as existing in normal embryonic tissue, will induce the formation of nerve tissue, it seemed useless to continue the search for any one substance as the organizer. This led Waddington to suggest that the place to study differentiation was rather in the 'competence' of the reacting tissue which actually carries out the differentiation.

This may remind one of that other situation when in 1913, the discovery was made that embryonic development in lower vertebrate ova could be initiated artificially without spermatozoa, in fact that parthogenesis, virgin birth, could be induced by a pinprick, heat or other mechanical means.

We must now refer to the developments in the study of heredity that took place at the same time. The rediscovery of Mendel's laws was followed fairly quickly by the realization that the individual hereditary factors that Mendel had stipulated were located in the chromosomes of the cell nucleus. In 1909, the term 'gene' was coined by Johannsen to denote these hereditary factors. Massive and thorough breeding experiments, especially with rapidly reproducing small insects, demonstrated the powerful working of the genes.

The considerable degree of variation produced in sexual

propagation through the factor of meiosis, which is the reduction of the chromosomes to half the number in each germ cell before fertilization, creating vast possibilities of combinations, was discovered. Artificially induced mutations were achieved by irradiation of chromosomes in germ-cells. Ultimately, individual genes became manipulable. Darwin's theory of evolution by natural selection seemed to be ascertained and nature became not only comprehensible, but possibly also controllable. The principle of heredity was grasped and could now almost be mastered. Yet the real breakthrough in genetics was to come only in the second half of the present century.

In 1935, O. E. Schotte of Germany carried out an elegant transplant experiment that beautifully illustrates the peculiar interaction between organizer and genes in embryonic development. From the underside of a frog embryo, he removed a slice of cells which would normally have become flank skin and transplanted it into the prospective mouth area of a salamander embryo. Frog and salamander, although both amphibia, have very obvious differences in structure. The salamander has teeth in its jaws, and on each side of its mouth stabilizers for swimming. The frog tadpole has no teeth, but horny jaws and suckers on either side of the mouth.

In Schotte's experiment, the transplanted flank skin of the frog formed a jaw in the salamander but without teeth and with suckers instead of stabilizers; thus the salamander organizer prompted the frog cells to form a mouth, but the frog genes carried the specific form qualities with them. From this we can conclude that the organiser determines the general, overall structure of an organism, its form or gestalt, but that the genes determine the specific, hereditary details of the organs and tissues.

The general attitude of biologists to preformation and epigenesis in the first half of the twentieth century was very clearly formulated by Huxley and de Beer (1934, 2): 'The modern view is rigorously preformationist as regards the hereditary constitution of an organism, but rigorously epigenetic as regards its embryological development.'

4. THE BEGINNINGS OF EXPERIMENTAL EMBRYOLOGY

The first edition of Hamilton and Boyd's excellent book, *Human Embryology*, (1945, 2) contains an interesting statement on heredity and environment, which was changed in later editions:

> It is now well established that the special characteristics of an organism (e.g., colour of skin, eyes and hair, hair type) are due to its nuclear genic equipment. The more general characteristics (e.g., those enabling us to classify it as a man or a chimpanzee, as a primate or a carnivore, as a mammal or a reptile) are controlled by factors which are not yet understood.

Schotte's countless experiments demonstrate and establish the fact that the differences in families, orders and classes as well as individual differences of organisms are governed by genetic principles, whereas the overall general form-principle is not genetically pre-determined, but arises epigenetically in ways not yet fully grasped.

5. The discovery of genes

By the middle of the twentieth century it was realized that genes and chromosomes were essentially of the nature of nucleic acids, and finally it was established that the gene consisted of deoxyribonucleic acid, DNA, and of ribonucleic acid, RNA. The extraordinary power of DNA was demonstrated with viruses and bacteria. Thus virus DNA invading the bacteria produced, from the substance of the bacteria, complete copies of itself, revealing that it carried the secret of the hereditary transfer mechanism. But how was it actually done?

In 1953, Watson and Crick devised a physical model of the DNA molecule. The sugars and phosphates of the molecule arranged in continuous alternation were envisaged as two long twisted chains, forming what was called a 'double helix'. The two spiral chains of the double helix were interconnected at all their sugar-phosphate links by four different nitrogen compounds or bases, creating a kind of twisted rope ladder from the double helix. The four different bases, linking only in pre-determined pairs, present practically unlimited numbers of combinations in their sequence as rungs on the ladder. Basically, the idea was that the splitting of the rope ladder of the double helix by unzipping the countless rungs of pairs of bases would set the pattern for an identical reproduction of the chain, by attracting free bases from adjoining material in corresponding numbers, type and sequences, thus providing the mechanism for the transmission of genetic information.

The model of the double helix did not only answer all the requirements for an explanation of heredity, but was also so ingenious and beautiful a creation that its fame spread far beyond the confines of the scientific world to the arts, commerce and social life. For instance, the beautiful French exhibition *L'human-itée* which took place a few years ago, had as its sign on posters, a double helix formed of gold bars.

In the sixties and seventies of the present century, a number of as yet unexplained inborn diseases (not mongolism) were described and linked to chromosomal pathologies, most of them of a hereditary nature, arising anew from unknown causes in each case. Among the newer developments in cytogenetics are hybridization attempts through which genes were transferable from one species to another. With this there is, for instance, a hope that a harmless bacterium would be 'programmed' to produce insulin by the implantation of a gene, and that diabetics could be cured by being infected with this bacterium.

A further important field of research lies in the control of the interaction between genes and enzymes. All these developments in biology have extraordinary practical implications for medicine and for plant and animal breeding in particular, and open up so far unimaginable possibilities of control and power.

Understandably, this has resulted in a general overrating of the significance of heredity and in the popular belief that embryogenesis can now be explained genetically. Dawkin's brilliant and entertaining book, *The Selfish Gene*, is a good example of what is really only a half-truth. Even Hamilton and Mossmann have altered their passage on epigenesis and preformation in the 1972 edition of their *Human Embryology*. This passage in the first edition (1942, 2) reads:

> The embryological investigations of the past hundred years have demonstrated most conclusively that the actual processes of development are of an epigenetic nature but the doctrine of preformation has been reintroduced, in a much modified form, in the explanation of the facts established by modern genetics.

The corresponding passage in the fourth edition (1972, 6) reads:

> The embryological investigations of the past hundred years have demonstrated that the actual processes of development are of an epigenetic nature . . . Modern genetics, however, has shown that the genes located in the

chromosomes of the nucleus of the zygote [the fertilized ovum] carry the necessary information enabling normal development to occur. The hereditary constitution is therefore determined by the panoply of genes in the chromosomes and, in this sense, ontogeny is essentially a gradual revelation of the plan stored in the *genome* (the collective term for all genes).

While this statement on modern genetics goes well with popular opinion to which it has probably largely contributed, it is factually not correct and has been disproved by scientific experiment. The genes in the zygote do carry the necessary information to individualize embryogenesis on a hereditary basis (occasionally causing abnormality), but that they do not carry the necessary information for normal development has been dramatically demonstrated by repeated attempts to grow normal zygotes in a test-tube; this was successful inasmuch as growth of different tissues was concerned, but it has never produced normal development, never an axial skeleton, never a formed-out organism, but always only a formless mass of teratosis (monster formation).

While it is true that the hereditary constitution of an organism is determined by the panoply of genes in the chromosomes, ontogeny is not a gradual revelation of a plan stored in the genome as there is no such 'plan' in the genome, but only a vast amount of specializing and distinguishing detail. Modern genetics provides an adequate and lucid explanation of the vast number of distinguishing features in the millions of human beings (with the exception of identical twins), but not of the human form or gestalt. It explains the immense variety of features in the families, orders and classes of vertebrate animals, but does not account for the overall plan or form and structure that makes them all vertebrates. The plan which gives the form, or gestalt, has as yet not been fully grasped.

6. The search for the origin of form

In this chapter, we wish to refer to some trends of the contemporary search for a new understanding of the epigenetic nature of embryogenesis. The anatomist, E. Blechschmidt of Göttingen, developed a theory of form-processes in embryogeny by including the time-factor. He holds that the formative processes are a result of developmental movements of organs, cell formations, ultimately of the molecules in the cells. In the living organism, these developmental movements are regarded as movements against resistance. Here, the cell-plan is seen as active because it is adaptive, whereas the nucleus is relatively passive as the carrier of heredity. This kinetic polarity is taken further by the growth potential of the cytoplasm being resisted by the tension in the cell membrane. In the actual development of the overall form of the embryo, a corresponding tension between the embryo's metabolic growth forces and the space-restricting uterus plays an important role. (We may be reminded of Democritos' idea of the shaping factor of the womb in embryogeny referred to in Chapter 2.)

Growth is regarded by Blechschmidt as the stimulation from outside through nourishment. Development is regarded as a response to external stimuli, the genes safeguarding against any threat to the hereditary uniqueness of the organism, but not initiating differentiation. Blechschmidt even formed the idea that all later adult functions, including the psychological, were preceded by and rooted in corresponding growth functions in relevant metabolic fields of growth in the embryo.

Each organ according to Blechschmidt unfolds its structure and form-development in relation to its site in the organism. This means that all organs would have from the start, as parts of the organism, a form-determining function. This elementary function has to be seen primarily as developmental dynamic. Each organ acts in the framework of its formative functions according to the qualities it has achieved up to its respective phase of development. It functions always within its special circumstances.

Blechschmidt's work is built upon a massive and meticulous morphological study of embryonic development based on countless microscopic and macroscopic sections of embryos, which he reassembled in enlarged plastic models, producing a unique exhibition of successive stages of embryo forms and anatomy.

In 1968, Arthur Koestler organized a symposium entitled 'Beyond Reductionism'. It provided a platform for sixteen scientists to share ideas on how to overcome the prevalent reductionist belief in science, that all human phenomena can be reduced to elementary responses observable in animals, and that these responses in turn can be reduced to physical-chemical laws.

The reductionist assumption derives from the mechanistic world view of the nineteenth century, and was abandoned by physics early in the present century, whereas it still prevails in the life-sciences from genetics to psychology. By denying a place to values, meaning and purpose in the interplay of blind forces, reductionism has cast its influence beyond the confines of science over the wider cultural climate of our time.

Several of the contributions made in this symposium are of considerable relevance to embryogenesis. Paul A. Weiss of the Rockefeller University, New York, argued that the principle of hierarchical order in living organisms reveals itself as a demonstrable descriptive fact, regardless of the philosophical connotations it may imply, and that the necessity becomes compelling to accept organic entities as systems subject to network dynamics in the sense of modern systems theory, rather than as bundles of micro-precisely programmed linear chain reactions.

Paul Weiss supported his case by demonstrating with electron-microscopic and lower power photographs of parts of cells and multi-cellular organs, that their highly intricate and elaborate patterns and forms did not necessarily coincide with either molecular or cellular structures, but were obviously determined by superimposed hierarchical systems of structuring. He held that (Koestler and Smithies 1969, 37–39):

> . . . genes, highly organized in themselves, do not impart higher order upon orderless milieu by ordainment, but . . . they themselves are part and parcel of an ordered system, in which they are enclosed and with the patterned dynamics of which they interact. The organization of this supra-genic system, the organism, does not even originate in our time by 'spontaneous generation'; it has been ever present since the primordial living systems, passed down in uninterrupted continuity from generation to generation through the organic matrix in which the genome is encased. The organization of this continuum is a paradigm of hierarchical order . . .
>
> In conclusion, genetics alone can account rather precisely for the differences between attributes of systems . . . but for a complete definition of the integral subjects which carry and display those attributes, genetics must call on *systems dynamics* for supplementation; . . . The dynamism of organization is *dualistic* . . . The loss of these integrative dynamics is the mark of death.

In the discussion on Paul Weiss's contributions to the symposium, David McNeill, professor of psychology at the University of Chicago, pointed out how the issue Weiss raised for living organisms is encountered equally in language. Previously, it had often been assumed that sentences were collections of elementary 'particles', words, and that the characteristics of sentences could therefore be understood by adding the parts.

Arthur Koestler's contribution to the symposium was called 'Beyond Atomism and Holism' in which he developed a system-theoretical model of 'Self-regulating Open Hierarchic Order' (SOHO). He brought this as an alternative to the rejected stimulus-response model of linear causation derived from classical mechanics. Some general features of this system are: the organism is regarded as a multi-level hierarchy of semi-autonomous sub-wholes, branching into sub-wholes of lower order and so on. Sub-wholes on any level of the hierarchy are referred to as 'holons'. Hierarchies are dissectible into their constituent branches on which the holons form the modes; the branching lines represent the channels of communication and control. Functional holons are governed by fixed rules and display more or less flexible strategies. Every holon has the dual tendency to preserve and assert its individuality as a quasi-autonomous whole; and to function as an integrated part of an existing or evolving larger whole. This polarity between the self-assertive and the integrative tendencies is inherent in the concept of hierarchic order and a universal characteristic of life. The self-assertive tendencies are the dynamic expression of the holon's wholeness, and the integrative tendencies of its partness.

These are only a very few features of the many dozens described by Koestler, which in their totality make his theory applicable not only in the field of biology, but equally in that of psychology and sociology, and aim to provide an interdisciplinary, general model.

C. H. Waddington, at the time professor of genetics in Edinburgh, brought important criticism and far-reaching and fascinating new ideas to the present theory of evolution, still rooted in Darwin's and Mendel's work and often summed up in the classical phrase, 'random mutation and natural selection'. The neo-Darwinistic theory of evolution has gained its wide scientific acceptance on the basis of Haldane's, Fisher's and Wright's mathematical support of the feasibility of evolution seen as mutation and selection.

Waddington pointed out that natural selection works of necessity on phenotypes, the actual qualities and features of individual organisms, but not on their genetic potential or genotype. According to Mendel's laws, these, too, can differ fundamentally because of the phenomenon of dominant and recessive characteristics. However, all three theories commit the basic error in logic of attaching coefficients of selection not to phenotypes but to genotypes, thus fundamentally weakening the whole scheme.

Waddington then showed that the logical, mathematical characteristics of genetic transmission and even of natural selection can be found in minerals, and gave as an example the continuing formation of surface dislocations in the growth of crystals. This reveals the inadequacy of a widespread definition of life as essentially characterized by genetic transmission and natural selection.

Waddington went on to show that in organisms such as viruses and bacteria with their thousands of genes, phenotypes may not differ too greatly from genotypes. These become more radical in differentiated organisms where the possible combinations of hundreds of thousands of genes with one another and their interactions with surrounding influences must provide truly multitudinous possibilities.

He then developed an image of a series of multi-dimensional spaces in which a genotype-space moves over into a phenotype-space through an 'epigenetic space', represented by vectors which tend to push the developmental processes into one direction or another.

Not all the epigenetic factors arise from instructions in the genotype; some of them originate in the environment. The same genotype, therefore, can produce a number of phenotypes according to the environment of the development system. Thus there is an essential indeterminacy in the relation between genotype and phenotype, the relation becoming determinate only if the environment is also taken into account.

Waddington emphasized the special kind of stability that the epigenetic space exhibits. It does not maintain a constant value, but rather a particular course of change in time. It leads to an ultimate, in some way predetermined, form through an ability to change its course or path. Waddington speaks of a 'homeorhetic system' which reacts to any environmental influence or impact, not by returning to the state in which it had been before, but rather coming back to where it would normally have arrived in its development some time later, somewhat like a river, forced out of its valley by a landslide, which does not return to the stream bed where the diversion occurred, but further down the valley.

Waddington created a new word for this stabilized time-trajectory, 'chreod' derived from the Greek *chre* ('it is fated' or 'necessary') and *hodos*, (the 'path'). He says that the epigenetic space is essentially a space of such chreods, which absorb and become defined by the genetic instructions from the genotype space, leading over to the phenotype space.

The increasing complexity of the epigenetic processes through which the genotype develops into the phenotype mitigates the randomness in mutation. To quote Waddington:

> In highly evolved, complex organisms the randomness of the basic gene mutation is, as it were, buried deep in the complexity. It is rather like the randomness in the shape of the pebbles which form the aggregate of the concrete out of which a bridge has been built. It remains true enough to say that the ultimate units, the pebbles in the concrete or the genes in the organism, have been produced by random processes, but this is almost irrelevant to the engineering of the bridge, and in many cases not much more relevant to the anatomical or physiological construction of the organism.

The question arises: What then is responsible for the engineering of the bridge, for the anatomical and physiological construction of the organism?

In a similar symposium to the Alpach Symposium referred to here, Waddington very tentatively suggested that something like the principle of Chladny's vibration phenomenon could possibly play a part in the form construction of organisms in the epigenetic space. He referred to the work of Hans Jenny, a physician and researcher in Dornach, Switzerland. Jenny had made a lifelong study of the forms arising in different finely granulated or powdered substances when exposed to various acoustic vibrations. His work had gone unnoticed for several decades, but suddenly reached the limelight when the Institute of Contemporary Art in London exhibited his work.

Jenny photographed and filmed his experiments in such a beautiful way that they presented a fascinating display of art at the I.C.A. and for the first time, became widely known, stirring some interest in scientific circles.

In the exhibition of Jenny's work, a film was shown among many other captivating exhibits, that appeared to depict a battle between two armies of medieval knights in heavy armour, one army on horseback, the other on foot. Factually, it was a film Jenny took of two different metal dusts exposed to the vibrations of two different pieces of music.

Many other examples were shown which demonstrated the highly differentiated, beautiful and complex forms that can arise when masses of small particles are exposed to simple or more complex sound vibrations. It was to this phenomenon that Waddington referred in connection with the question as to how form might arise in the epigenetic field. This may remind us of Harvey's and Borelli's comparison in the nineteenth century of the generating and shaping power of the semen to a magnetic field.

Professor Brian Goodwin of the Open University, in an important article 'On Morphogenetic Fields' makes an interesting criticism of the neo-Darwinistic theory of living organisms. He first takes up the relation of genotype, phenotype and environment which we encountered in Waddington's work and then examines the assumption, so brilliantly expounded in Dawkin's *Selfish Gene*, that the properties of organisms are reducible to their genes. Goodwin falsifies (a term used by K. Popper) this assumption with two examples: the first, that surgical modification of the morphology of an organism (the unicellular ciliate Paramecium) results in the transmission of this modification to all its offspring without change in genotype; second, the mermaid's cap, a green unicellular algae some three centimetres in length with a differentiated parasol-like cap at one end and a branched, rootlike rhyzoid at the other, interconnected by a thin stalk. When cap and rhyzoid, the latter containing the nucleus, are both cut off, a new cap is generated, usually at the end of the stalk where the former cap had been, but often also at the other end where the rhyzoid had been. Rhyzoids are rarely regenerated, the nucleus never. This shows that the undifferentiated stalk in the absence of a nucleus, which means without contribution from the genes, can create the highly differentiated cap or rhyzoid.

Goodwin concludes from this example of regenerative capacity that the morphogenetic field is primarily a property of the cytoplasm. His main argument, however, is

that neo-Darwinism and molecular biology are unable to
provide a theory of development because they lack
concepts of spatial organization and of relationship between

whole and part, the organism being seen as a collection of atomic elements, whether molecules or cells are adapted characters, whose principle of integration remains undefined.

Goodwin sees the living organism primarily and essentially as of fieldlike nature and the space-time processes of the maintenance of organic form and function in the adult as a continuation of development.

A basic property of the organic field is that from a part the whole can be reconstituted. This part may be a limp stump, as in some amphibia; it may be a fragment of the adult organism as in hybrids . . . or it may be the egg, a part of the adult which develops into a new organism after fertilization, the usual manner of sexual reproduction.

Goodwin makes an even more intriguing contribution in an earlier article 'On some Relationships between Embryogenesis and Cognition' touching upon language. In connection with a cellular slime-mould and with a hydroid, he first describes the continuity in embryogenesis and behaviour (also stressed by Blechschmidt), seeing both 'the processes of morphogenetic sculpturing and behavioural structuring as manifestations of the spatial-temporal order that arises from periodic wave propagation over an excitable continuum.' He describes embryogenesis as highly context-sensitive rather than as a stimulus-response mechanism; as very different stimuli may call forth very different responses in differing situations. The multi-potent quality of the embryonic cell is stressed with its variety of pathways or trajectories. Goodwin then states that developing cells have a degree of freedom; they can start to differentiate in one direction and then change direction later if this becomes appropriate, which means that complete field-instructions are not always present. It would seem that there is a kind of 'groping'. Developmental fields induce cells to adopt particular hypotheses concerning their prospective pathways. The cell then tests a hypothesis by a process of interaction with its neighbours, rather as we form our theories.

In addition, Goodwin points to the similarity between embryo-genesis and language as McNeill had done previously. He maintains that it is the context-sensitivity that embryogenesis and language have in common. In the sentence 'I saw the dog running', the meaning of the word 'saw' is only clarified by the context (you could also say 'I *saw* logs'). Goodwin then refers to the phenomenon of grammatically correct sentences that have no meaning in reality, such as 'The apple ate the boy', and demonstrates that the same thing can occur in developmental processes, for instance, in extro-gastrulation, which while consistent with the relevant developmental constraints, nevertheless constitutes a *cul de sac*.

The issues we have attempted to discuss here are very fully taken up in Rupert Sheldrake's *A New Science of Life*. This book sets out the well-established phenomena of morphogenesis that cannot be adequately explained by the neo-Darwinian mechanistic theory. Sheldrake also briefly describes and discounts the vitalist theory and, taking up the idea of morphogenetic fields from Waddington, he develops his 'hypothesis of formative causation'. He invests the concept of morphogenetic and motor fields with a physical reality, though stating that they work across space and time, influencing both form-development as well as the behaviour of living organisms through 'morphic resonance': 'The characteristic form of a given morphic unit is determined by the form of previous similar systems which act upon it across time and space by a process called morphic resonance.' No attempt is made to explain the origin of these fields, but the fields are moulded by the form and behaviour of past organisms of the species.

Sheldrake describes how the hypothesis of formative causation can be tested experimentally, and mentions certain experiments carried out in the past which can be cited in support of his theory. In these experiments, it appeared that after rats had been taught new patterns of behaviour, similar rats at a later time and in far-removed places would learn the same patterns of behaviour more easily and rapidly. This phenomenon has been documented in Harvard, Edinburgh and Melbourne in the course of the past fifty years.

While Sheldrake presents his hypothesis of formative causation as a scientific hypothesis in the sense of Popper's testability, he is aware of the metaphysical implications of introducing a new concept of formative causation between conscious causation in experience and energetic causation in physics. It may be difficult to accept this as physically causative as it is assumed not to be attenuated by space and time. The book, however, is not only remarkable for propounding a new scientific theory, but especially for its relevance to the relation between science and metaphysics of mythos. We shall take up this theme in Chapters 9 and 10.

A uniquely different approach to our subject which could possibly become one of decisive importance is that of Lawrence Edwards. His work is based on projective geometry, a branch of mathematics originating in the sixteenth century, possibly in connection with Leonardo da Vinci's and Dürer's studies in perspective. Among many others, Arthur Cayley perfected the algebraic methods of projective geometry in the nineteenth century which were then translated into pure geometry by Felix Klein at its turn. Briefly and simply put, projective geometry is a geometry of points, lines and planes which, with the inclusion of infinity, displays principles of polarity and trinity in the interaction of its elements.

For years, Lawrence Edwards has used the principles of projective geometry to show relationships between plant buds and seed forms. More recently, he has demonstrated that projective geometrical transformations between vortex forms and averaged forms of the human uterus create two-dimensional projections of successive forms of early human embryonic development. The same set of transformations provides a fully three-dimensional model of the general gesture which the embryo makes in forming the neural canal across the neural plate immediately after the disappearance of the primitive streak. (As yet unpublished personal communication).

Edwards' work may well lead to the first successful realization of Marcus Marti's striking idea of radiating beams of light as sources of form-development in embryogenesis and of Descartes's and Gossendi's premature attempts in their iatromathematical school, basing embryogeny on purely mathematical and geometric principles, which to some extent reappeared in the work of D'Arcy Thompson in this century.

From Science
to Mythos

7. Embryogenesis and Genesis

In the first two chapters of this book which presented a historic account of the changing thinking on embryogenesis throughout the centuries, I intended to show how this thinking wrested itself away from mythos and signification to causality and necessity, finally attempting to include new aspects of signification in the clarity of scientific thinking achieved in our own century. At the beginning of this century a similar process had begun in the field of physics, when Einstein with his theory of relativity and Planck with his quantum theory showed that in the area of sub-atomic particles neither the principle of causality nor that of objectivity could be absolutely upheld. The idea of time as a one-directional stream and that of inescapably three-dimensional space were determined. The idea that a cause must precede an effect and that their relation was invariant became untenable. In fact, the idea of necessity had to be replaced by that of probability, and the static idea of mass by that of dynamic processes.

This development in modern physics and its relation to ancient theories of significance is well and lucidly described by Fritjof Capra in *The Tao of Physics*. In modern biology, similar attempts have been made to look beyond the principle of cause-and-effect as we saw in the last chapter. But while the change in modern physics is universally accepted in university circles and even often referred to in ordinary education, this is not the case in biology. Nearly all universities maintain a strictly neo-Darwinistic outlook, although to do so is no longer scientifically completely justifiable.

Primary and secondary education identify embryogenesis largely with genetics and give the growing child and adolescent a one-sided picture, a half-truth, often presenting aspects of modern genetics in attractive forms, but altogether omitting the aspect of form development. Earlier on we mentioned Dawkins's *The Selfish Gene* which brilliantly and entertainingly describes the genetic half of the truth while leaving out the other half of form development, and we also mentioned Watson and Crick's 'double helix' depicted in gold bars as the poster for the French exhibition *L'humanitée*.

One wonders what is at the root of this misinterpretation and misunderstanding of the facts of embryology actually known today. An important part is probably played by the fact that modern genetics has opened up hitherto unforeseeable possibilities of practical development and control in medicine and in plant and animal breeding as well as in other fields, as we have earlier on suggested. It is for this reason that most of the scientific research in biology has concentrated on genetics, so that for any one scientist working in the field of embryogenesis, there may be several hundreds working in genetics.

A probably more relevant factor in this one-sided emphasis is the generally materialistic outlook of today which finds reassurance and security in the idea of causality and necessity in genetics. Yet we may ask whence the materialistic outlook comes. Has not just the half-truth of genetic preformation, of man being identical with his genes, sadly lowered man's self-evaluation and largely contributed to his materialism? Causality and necessity exclude freedom, dignity and values. Joseph Needham in his *History of Embryology* sees the expulsion of ethics from biology and embryology as one of the lasting achievements of many centuries of development: 'that good and bad, noble and ignoble, beautiful and ugly, honorable and dishonorable are not terms with a biological meaning' formulates an attitude that is generally accepted today.

Goethe would not have accepted such an attitude. He included the beautiful and the good, suffering and joy even in his studies on colour. The often-mentioned link between embryogenesis and language also raises the question of meaning and ethics. Meanwhile, however, we shall assume that ethics and aesthetics are rightly kept out of biology and shall rather query the present-day definition of mythos in its relation to science.

A *myth* is usually defined as a fictitious legend accepted as historical and embodying the beliefs of a people. A myth is not only regarded as incompatible with science, but as an antithesis, the word 'myth' frequently being used to denote a fancy or untruth. Yet earlier on we showed how science developed out of myths. Modern anthropology describes how ancient peoples lived as successfully under the guidance of their myths as men live now under the aegis of science. We shall return to these questions later and shall meanwhile take it that ethical and aesthetic considerations, out-of-bounds as they are for science, are acceptable in myth where causality and necessity are secondary. In the final chapters, we shall argue that the central factor in this issue is an erroneous evaluation of the nature and essence of science, particularly in its relation to mythos.

In this chapter, an attempt will be made to develop a new and contemporary understanding of mythos based on observation of early embryogenesis. The question will not be posed as to how embryonic development comes about or what the forces are that fashion the embryo, but rather we shall pose the question as to what meaning, what significance we can experience when we accept modern scientific descriptions of early embryonic development as images, or imaginatively.

When we look at a picture in an art gallery, we do not inquire as to the chemical composition of the paints and the canvas, but we allow our imagination free rein in the experience of what the picture sets off in our minds. In other words, we do not aspire to objectivity, but rather surrender our entire person as a resonant instrument to what impresses us. A similar stance will be employed here.

Human embryogenesis is seen today in three phases. The first phase covers just under three weeks and comprises the development of the embryonic membranes from the fertilized ovum, no recognizable human form emerging during this time. The second phase, lasting forty days up to the end of the second month, is the embryonic phase proper in which the embryo gradually develops its human form with head, trunk and limbs and all its organs, although it is just over one inch in length. The third and last phase covering the remaining seven months up to parturition,

the foetal phase, is mainly one of growth and further differentiation. It is in the first two phases that the decisive events of
embryogeny take place, but to begin with, we shall discuss the
first phase only.

It is commonly stated that the first phase of human embryogeny
begins with the fertilization of the ovum by the sperm, initiating
cell division and ending with the formation of the embryonic disc
with its three germ layers from which later the three types of
body tissue will develop, as well as establishing the three spatial
axes of the embryonic form. The exact duration of this phase
is difficult to ascertain, because with the exception of *in vitro*
fertilization, the actual moment of union between ovum and sperm
cannot be observed. The figures given vary from seventeen to
twenty days.

Most textbooks divide this first phase into six stages, albeit
stressing the fact that the transitions are fluid and development
continuous. It is interesting to note that while human embryogeny
is specific from the first moment of development, in fact right
from the ovum, it is in the first phase in which no embryo form
appears, that in its six stages it has been likened to the otherwise
very different six stages of metazoan (animal) embryological
development in its entirety. The metazoan ovum develops directly
through morula, blastula, gastrula, neurula into its embryo form.
The human embryo, however, begins to develop only after the
first phase of seventeen to twenty days has passed and the four
membranes — amnion, yolk-sac, allantois and chorion have been
developed.

We shall now describe each of the six stages of the first phase
of human embryogeny according to various textbooks and actual
photographs. Later we shall relate them to the six days of creation
as described in Genesis as well as to medieval miniatures illustrating these six days.

The signum of the first of the six stages of embryonic development
is the dual impact of polarization and division. It is introduced by
the development of the ovum and sperm respectively from female
and male primordial germ cells of about the same size. Both go
through a twofold reduction-division in halving their chromosome

complements. The spermatogonia go through spermatocyt and spermatid stages, separating their X and Y chromosomes, each germ cell ultimately producing four spermatozoa, two with X and two with Y chromosomes. In this process, most of the cytoplasm is lost so that in the end, the sperm consists almost entirely of nuclear substance. The whole way from relatively large spermatogonium to minute sperm is centripetally directed towards the lumen of the spermatic tubules which make up the testes.

From the female germ cell, the oogonium, only one mature ovum arises from the two reduction-divisions, the other three cells being discarded as so-called polar bodies containing almost only nuclear substance. The oogonia lie amassed near the periphery of the ovaries, surrounded by follicular cells. The maturing of the ovum is accompanied by follicular growth, the follicle gradually filling with fluid. This finally leads to the expulsion of the mature ovum from the ovarian follicle outward into the abdominal cavity. The ovum is by far the largest cell in the human body — over one tenth of a millimetre in diameter — and can be seen with the naked eye.

The ovum is about a hundred thousand times larger in volume than the sperm and for any one ovum, two to three hundred million sperm will vie. While the ovum is the precursor of all subsequent development, it is nevertheless regarded as the least specialized cell type, holding the potential of differentiating form-creation in its huge cytoplasm like a void, while the directions for individuation are concentrated in the relatively small nucleus.

The ovum is completely round and enclosed in a highly light-reflective thick membrane, the *zona pellucida*, and surrounded by a bright halo of follicular cells known as the radiating crown or the *corona radiata*. For a moment, the ovum hovers in suspense before being taken up by the wafting finger-like fimbria of one of the Fallopian tubes. Semen having been introduced into the vagina, millions of spermatozoa travel in a great stream upwards through the uterus and into the Fallopian tubes.

The spermatozoon is the smallest cell in the body, only 0.01 to 0.02 mm in diameter, and consists almost solely of a small torpedo-shaped nucleus and a fine flagellating tail. The head of the spermatozoon is highly light-reflective so that the rapidly oscillating streams of sperm rushing towards the ovum appear like a flash of lightning when seen under the microscope. Fertilization occurs through the penetration of the ovum by one spermatozoon which sheds its tail, enlarges its head, and fuses with the nucleus of the ovum.

A complicated, chaoticizing and restructuring process follows which leads ultimately to cleavage, the division of nucleus as well as cytoplasm into two cells, then to four, later to eight and sixteen cells, continuing until the end of the first stage is reached in the formation of a cluster of cells, the so-called morula. In spite of the many cells the morula contains, it is not any larger than the original ovum.

This cell division can be regarded not only as a quantitative process leading from one large fertilised ovum to a cluster of smaller, similar cells, but also as a qualitative division of those cells which will continue to differentiate ever further into all the many tissues and organs from those other cells which will retain the general potential of procreation of the whole embryo. The former group consisting of the cells which will form bone, nerve and the other tissues as well as liver, heart, lung and other organs, are referred to as *soma*, meaning 'body' as distinct from the germ cells. Another aspect of the qualitative division is the separation into those cells that later form the embryonic sheaths or membranes.

After the dramatic event of the first cleavage, there is a pause; each subsequent division is again followed by a phase of rest. The whole process of polarization and division from the first cleavage to the many-celled morula takes about four to five days (Figure 1).

In Genesis (1:1–5) the first day of creation is described as follows:
    In the beginning God created the heavens and the earth [the
    polarization of female and male germ cells into ovum and
    spermatozoon respectively]. The earth was without form and
    void and darkness was upon the face of the deep; and the
    Spirit of God was moving over the face of the waters
    [hovering suspense of the ovum void of form; the radiating
    crown of the Spirit of God moving upon the face of the
    waters].

7. EMBRYOGENESIS AND GENESIS

> And God said, 'Let there be light'; and there was light
> [the appearance of the lightning stream of sperm]. And
> God saw that the light was good; and God separated the
> light from darkness [cleavage, division]. God called the
> light Day, and the darkness he called Night. And there was
> evening and there was morning, one day [the light of the
> sperm starting off the division into day and night].

However, day and night as such are only created later with the
creation of sun and moon. The Hebrew words used here for
heaven and earth are *hashamayim* and *ha'aretz*. *Hashamayim* can
be translated as 'that which steps out into the light of appearance'
or that which reveals itself, which forms the body, the *soma*.
Ha'aretz is the dark earth, that which retains the quivering potential of generation, procreation, germ substance.

There is a medieval miniature that illustrates the first day of
creation beautifully (Figure 2). God stands in front of a golden
background surrounded by angels and devils. He nearly fills the
entire space. In one hand, he raises up a shining white sphere
which is inscribed by a right angle of vertical and horizontal lines;
hashamayim — that which steps forward into the light — the
soma. The other hand of God, directed downwards, holds a dark
unformed lump looking perhaps like a large piece of coal (the
Philosopher's Stone?), slightly suggestive of a pentagonal form;
ha'aretz — the dark quivering potential of procreation.

The biblical text concerning the second day of creation may strike
one as highly perplexing (Gen.1:6–8):

> And God said, 'Let there be a firmament in the midst of the
> waters, and let it separate the waters from the waters.' And
> God made the firmament and separated the waters which
> were under the firmament from the waters which were
> above the firmament. And it was so. And God called the
> firmament Heaven. And there was evening and there was
> morning, a second day.

How can one possibly imagine a firmament, a heaven in the midst
of waters, dividing waters above from waters below?

And yet exactly this appears in the second stage of the first
phase of early embryonic development (Figure 3). By about the

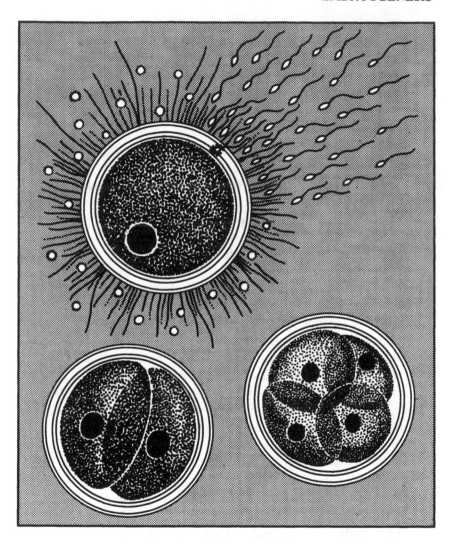

Figure 1. First stage. Fertilization of ovum by sperm; division and morula formation.

Figure 2. The first day: And God said, 'Let there be light'; and there was light. And God saw that the light was good; and God seperated the light from the darkness. (Illustration from Old Testament Miniatures)

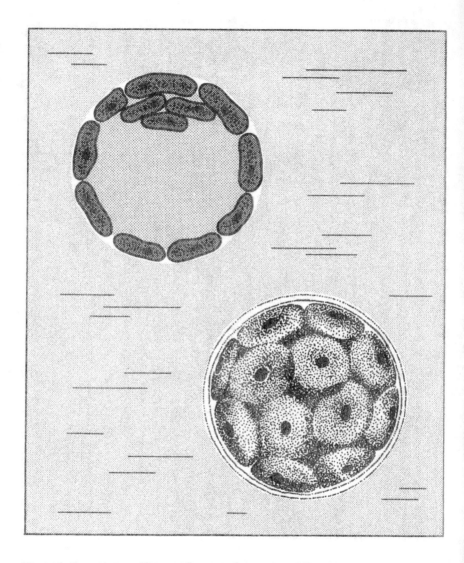

Figure 3. Second stage. The morula enters the womb and changes into the blastocyst; fluid inside and outside.

Figure 4. The second day: And God said, 'Let there be a firmament in the midst of the waters, and let it separate the waters from the waters.'

Figure 5. Third stage. Nidation. Beginning of gathering of amniotic fluid and forming the first germ cell layers within the mesenchyme.

Figure 6. The third day: And God said, 'Let the waters under the heavens be gathered together into one place, and let dry land appear.'

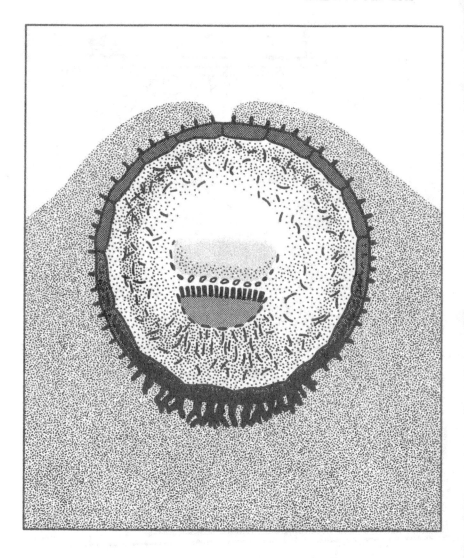

Figure 7. Fourth stage. Implantation. Forming of the starlike villi around the 'firmament' of the trophoblast. Beginning of the sunlike placenta and of the sickle-moon form of the body stalk and allantois.

Figure 8. The fourth day: And God said, 'Let there be lights in the firmament of the heavens to separate the day from the night ...'

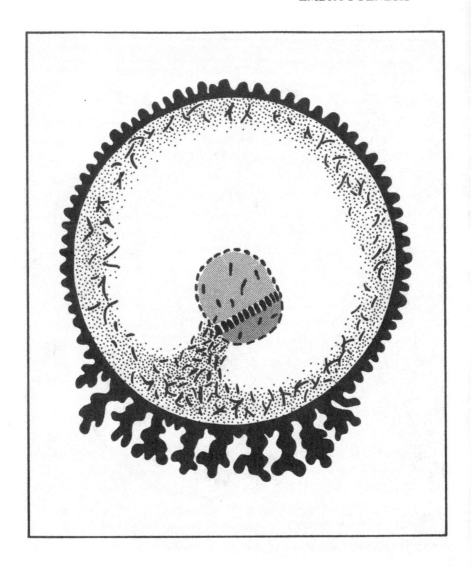

Figure 9. Fifth stage. Pulsating blood-islands appear on the amnion connecting the stalk and the yolk-sac.

Figure 10. The fifth day: And God said, 'Let the waters bring forth swarms of living creatures, and let birds fly above the earth across the firmament of the heavens.'

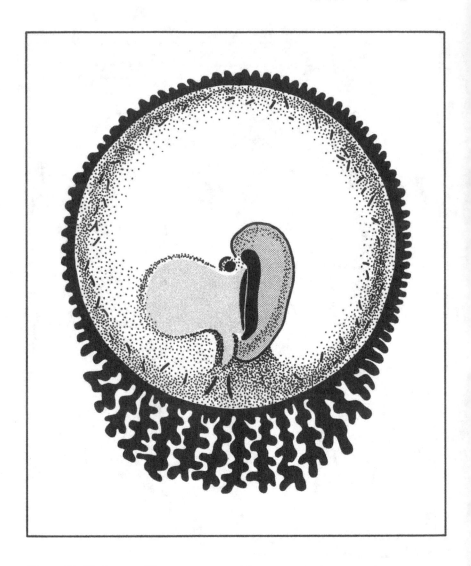

Figure 11. Sixth stage. First appearance of the embryonic disc out of the three germ layers. Beginning of the chorda, neural tube and heart formation, and intestinal invagination.

Figure 12. The sixth day: Then God said, 'Let us make man in our image, after our likeness ...'

fifth day after conception the morula, having moved down the Fallopian tube into the uterus, begins to change. While cell-division continues, fluid from the uterine cavity begins to seep through the pellucid membrane and the outer cell-layer of the morula. An inner cavity of fluid arises, the blastocoele; the pellucid membrane dissolves, the outer cells of the morula become flattened as they are pushed outwards and ultimately form a hollow sphere, the blastocyst. As a result of these changes, the inner cell-mass becomes attached eccentrically to the inner side of the outer layer of flattened cells which, because of the part they are going to play in embryonic nutrition, are called the trophoblast. For a short time, the blastocyst floats in the fluid of the uterine cavity and is also filled with the same fluid; a sphere, a firmament, dividing waters without or above from waters within or below.

This firmament as a sphere separating the waters is clearly depicted by the medieval miniaturist in his illustration of the second day of creation (Figure 4). God has stepped to the side, his face and right hand are lifted up to the deep blue of the background. His left hand rests on a large sphere formed by a wavy line very much like the flattened outer cells of the blastocyst. This sphere is filled with the same blue waters that form the background. Above and inside the sphere, opposite to what looks like a little hill below, clouds appear indicating the waters below the firmament.

The third stage of the first phase of the development of the human embryo is usually referred to as nidation (Figure 5). Between the seventh and tenth days, the trophoblast proliferates outwards, more so on the side of the inner cell-mass, thus partly losing its cell-structure. With this proliferating side, the blastocyst first attaches itself to the uterine wall which it then begins to invade until the uterine wall closes over it. During the process of nidation, the inner cell-mass begins to differentiate into two layers of dissimilarly formed cells, the first appearance of the ectoderm and, central to it, a second less distinct group of roundish, slightly flattened cells which will become the endoderm.

Between these dishlike layers and the trophoblast, there

appears a small accumulation of fluid in the amniotic cavity which separates a fluid area from the remaining solid cell-mass. As a second step in this stage, loose fibrous mesenchyme structures appear inside the blastocoele which, disappearing again, will give rise to the extra-embryonic coelom (or primary yolk-sac). Into this area, the yolk-sac will develop from the inner layer of the cell-mass.

We can readily recognize the developmental steps of this third stage in the words of Genesis which describe the third day of creation (1:9–11):

> And God said, 'Let the waters under the heaven be gathered together in one place [the amnion], and let dry land appear [the inner cell-mass of the embryonic disc].' And it was so. God called the dry land Earth, and the waters that were gathered together he called Seas. And God saw that it was good. And God said, 'Let the earth put forth vegetation, plants yielding seed, and fruit trees bearing fruit in which is their seed, each according to its kind, upon the earth.' And it was so.

Yet later, we read that 'God created . . . every plant of the field before it was in the earth and every herb in the field before it grew' (Gen.2:5). From these words, we may conclude that the plants created on the third day were rather the archetypal plant forms that preceeded the actual plants growing in the earth. Rudolf Steiner spoke of the appearance and disappearance of gigantic plant forms in the albuminoid atmosphere surrounding the earth in early phases of evolution. We can still see tiny traces of plant forms in a colloidal medium in the so-called moss agates in which various metal salts created plantlike forms dissolving in silica-gel before solidification.

Just as we can observe the separation of sea and land in the formation of the amniotic fluid in the differentiating inner cell-mass, so may we recognize the creation of the archetypal plant forms on the third day of creation in the fibrous mesenchymal growth in which cell boundaries appear and disappear.

Again we find this beautifully illustrated in the medieval miniatures depicting the third day (Figure 6). We see God standing once more on the left before a rose-coloured background, turning with face and hand towards the sphere of his creation, the tropho-blast. In the latter, the little hill of earth has moved to the left, allowing the green water of the sea to appear on the right. The rose-coloured atmosphere is filled with plant forms reaching up the clouds above.

By the beginning of the fourth stage which extends from about the eleventh to the thirteenth day, the yolk-sac has developed and the trophoblast begins to grow so that amnion and yolk-sac, separated by the embryonic disc, look relatively small; the uterine wall has completely closed over the implanted trophoblast (Figure 7). During this stage, the growing trophoblast becomes studded with tiny villi all over its surface. Over the part where the amnion is attached inside, the villi grow larger in a disclike area in prep-aration for the future placenta.

Corresponding to this development, chorionic mesoderm grows inside between the amniotic cavity and the trophoblast, forming the bodystalk which will later become the umbilical cord. In this development of the bodystalk, amnion and yolk-sac begin to turn anticlockwise so that the bodystalk connects where amnion and yolk-sac join at the embryonic disc. At this point, a small sickle-shaped extension grows from the yolk-sac into the bodystalk, the so-called allantoic bud.

We may imagine the larger villi which will develop into the life-giving disc of the placenta as the sun in the firmament of the trophoblast, now studded with the countless tiny villi as the stars. A little lower, between earth and firmament in the con-necting stalk, there appears the sickle of the moon, the allantoic bud.

In Genesis it says: 'And God said:. "Let there be lights in the firmament of the heavens. . .".' (1:14) Then (1:16): 'And God made the two great lights, the greater light to rule the day, and the lesser light to rule the night; he made the stars also . . . And God saw that it was good. And there was evening and there was morning, a fourth day.'

Here the miniaturist has returned to the original golden back-ground (Figure 8). God is again turned towards the sphere of the firmament which is now rose-coloured. Under it, between clouds and blue sky, a tufted villous sun is faced by a waxing sickle moon. Minute stars are just discernible in the blue heaven.

On the fifth day of creation, the animals in water and air are created (Figure 10), and in the beginning of the sixth day, the land animals. We see these events in the fifth stage of early embryogeny when towards the fourteenth and fifteenth days, pulsating blood islands appear, first in the bodystalk and around the allantois, then on the yolk-sac (Figure 9). Only later will these blood islands flow together and form the first blood vessels which will eventually link up with heart and with the placenta as the latter begin to develop. But at this early stage, the pulse of the blood in many independent islands, like separate living entities, appears long before the heart itself has begun to form. Movement of cells also occurs between a protuberance in the posterior ecto-dermal layer, the primitive streak, and the extra-embryonic mesoderm in the bodystalk. In connection with the movement of mesodermal cells in the periphery, ectodermal cells bud and grow into mesodermal cells which move downwards and spread out between ectoderm and endoderm.

It is only in the sixth and last stage of the first phase of embryonic development, however, that the trilaminar nature of the embry-onic disc and its three spatial axes becomes apparent (Figure 11). Already in the third stage around the eighth and ninth days, the embryonic disc had appeared, then consisting of two layers only; of ectoderm, the bottom of the amnion, and of endoderm, the top of the yolk-sac, forming a round disclike plate. From the fifteenth to the seventeenth day on, a third, middle layer, the mesoderm, appears between the first two in connection with the appearance of the so-called primitive streak. By that time, the embryonic disc has assumed an oblong shape with a wider, rounder cranial pole and a narrower, more pointed caudal end. The primitive streak appears as a furrow in the ectodermal layer, reaching from the caudal end to the middle of the embryonic disc where, beginning with the blastopore, the notochord, a tube of mesodermal cells, invaginates straight forwards between ectoderm and endoderm, marking out the longitudinal axis, the place where later the spinal column will develop. The mesoderm also spreads to both sides between the other two germ layers and to either side of the notochord, thus not only are up-and-down established between cranial and caudal ends by the chorda, but also dorsal and ventral sides by amnion and yolk-sac, and ectoderm and endoderm, as well as left and right by the latterly invading mesoderm forming the first somites.

Within this threefold spatial structure, the three germ layers will give rise to the three great organ systems: the ectoderm to the nervous system and outer skin, the endoderm to the intestines and most of the inner organs, and the mesoderm to the skeleton, muscles and blood as well as to that which gives form and movement.

Although during this last stage before the end of the third week, no recognizable human form appears, areas and gestures of form development can be discerned. A double ridge appears on the floor of the amnion, which, rising up, will close over the groove between, gradually forming the neural tube from which the nervous system will develop. Below and in front of the embryonic disc, where amnion and yolk-sac meet, a mesodermal cardiac plate and pre-cardiac cavity, which will later form the heart, become visible. From the endoderm, the top layer of the yolk-sac, cranial and caudal structures develop, the pharyngeal and cloacal membranes respectively, marking the areas where in the fourth week, the invagination of the yolk-sac will begin to form foregut and hindgut as the beginnings of the intestinal system; but still no human form has appeared.

Yet for the sixth day of creation, Genesis says (1:26–27):
Then God said, 'Let us make man in our image, after our likeness; and let them have dominion over the fish of the sea, and over the birds of the air, and over the cattle, and over all the earth, and over every creeping thing that creeps upon the earth.' So God created man in his own image, in the image of God he created him; male and female he created them.'
Our medieval miniaturist again presents us with a delightful picture (Figure 12). God is on the left of the sphere which contains the green sea with a whale, the red earth with horse, cow and

deer, lion and sheep, the air filled with a host of birds; sun and moon and stars in the blue heaven. On the side nearest to God, we see Adam who is being gently led out of the sphere by God who holds him by the arm. The miniature seems to illustrate human birth from out of the fourfold house that the four membranes — chorion (trophoblast), allantois, amnion and yolk-sac — have built during the six stages of early embryonic development.

While we have every reason to take delight in the contemplation of this gently but deeply satisfying picture, we have to remember that the sixth stage of early embryonic development ends with the seventeenth to twentieth day without the emergence of anything like the human form. Thus we may have to clarify what it is that is actually achieved by the end of the first phase. But before we attempt this, we shall first pose the question whether we (as well as the miniaturist) have rightly understood what is said in Genesis concerning man's creation in the image of God. Do we not tend to take this as meaning that God looks like a man, an oldish man with a beard, a kind of father-figure?

In connection with the third day of creation, we found it helpful to consult the second chapter of Genesis which begins with the words: 'Thus the heavens and the earth were finished, and all the host of them.' After telling of the seventh day, Genesis continues with the creation of the archetypal plants to which we referred earlier when it says: '. . . there was no man to till the ground . . . then the LORD God formed man out of dust from the ground, and breathed into his nostrils the breath of life; and man became a living being.' Only after this are the plants that grow on the earth and the actual living animals created. We assumed earlier on that the twofold creation of the plants indicates that the creation of the first chapter of Genesis is the creation of archetypes and that the creation described in the second chapter is the *actual* creation.

We shall now assume that the same principle applies to the creation of man. The first creation then would not be that of a living soul, but of an image of God; not of a man to till the ground, but of a man in God's likeness; not male only, but male and female — androgynous man. (That woman was made from a rib out of Adam's side would seem to be a somewhat patriarchal interpretation on the part of Moses. Thus although the future sex of the embryo is determined by the chromosome distribution at fertilization, the apparent secondary sex characteristics, appearing only at about the seventh week of embryogeny from an undifferentiated form, bear features rather of the female than the male.)

While the Jews were not permitted to make for themselves an image of God, their relation to God was certainly of that to a father, albeit a more vengeful and jealous father than the Christian Father God. But occult Jewish tradition had formed an image of the archetypal, androgynous man, Adam Kadmon, earlier referred to as Adam Protoplastes. This original, archetypal Man was envisaged within or *as* the zodiac, as the all-comprising, all-embracing periphery of the cosmos, perhaps as 'the generations of the heavens and of the earth when they were created', an image of the universal principle of form-creation in which the human form is revealed as the totality of all animal and plant forms, containing as its core the spatial and organic threefoldness inherent in all living organisms in the three spatial axes of cranial-caudal, dorsal-ventral and left-right, and in the three germ layers established by the end of the first phase of embryogeny.

This cosmic human form, however, also contains the universal spherical periphery in the fourfoldness of chorion, amnion, allan-tois and yolk-sac, representing four generations of the heavens and of the earth. Rudolf Steiner described these four generations of the evolution of the earth in their relation to the fourfold aspect of human spiritual nature in extraordinary detail in his book *Occult Science*.

K. König (1967), K. Appenzeller and F. Wilmar have described in varied and impressive ways the connection between elements of embryogeny, the biblical creation, and Steiner's ideas on evolution.

We should like to mention here an idea first put forward by König, suggesting that the four embryonic membranes become the respective seats of the four spiritual bodies or souls first postulated by Aristotle, then by Thomas Aquinas and later described by Steiner, who maintained that while the spiritual-physical form which he called the 'spirit germ' descends from the cosmos at the moment of conception and sets off form-development in the fertilized ovum, the individuality of the person to be born with his sentient and vegetative bodies only incarnates after the seventeenth to twentieth day, when the embryonic membranes have been prepared. M. Hoffmeister presents a collection of statements on this subject made by Steiner and others.

Our main concern now, however, is with the Man created in the image of God, this form-archetype, which may not only be the precursor of the second man, the man made from the dust of the ground, but also his creator.

Mythologically seen, it is the cosmic, zodiacal form of Man, the Adam Kadmon, that the descending 'spirit germ' brings down which gives rise to the threefold embryonic disc in its sheaths, and which will then bring about the human embryo form in the ensuing forty days in the second phase of embryonic development.

8. Embryonic form development and the cosmos

Reference to the zodiac today evokes in most people a sense of the suspect, particularly when the different zodiacal signs are regarded as having an influence on the individual lives of men. The twelve constellations that constitute the zodiac are neither the most conspicuous nor the most important of the many constellations of the starry sky. Like all constellations, those of the zodiac appear as visual phenomena, though the single stars composing the zodiacal images are not necessarily in any spatial proximity to one another, but are distributed over tremendous depths and widths between brighter and fainter stars.

The twelve constellations of the zodiac are significant in that they provide a relatively fixed and permanent scansion of the ecliptic and thus facilitate a direct perception of the ever-changing and dynamic relationships between the earth and the sun, the moon and the planets. Normally, we remain oblivious of these relationships, yet we experience in the course of the seasons, in the equinoxes and solstices, the ever-recurring 'dialogue' between sun, earth and stars.

Not only did the zodiac provide an instrument or measure by means of which the interrelated movements between the various heavenly bodies and the earth could be ascertained by the ancient Babylonians and Chaldeans, but it was the conduct of the sun itself, tracing its course over the skies, that brought the zodiac into the consciousness of those ancient observers of very early pre-Christian times. Those men, possessing powers of experience, intuition and imagination long since lost and today replaced by science and technology, gave names to the constellations the sun passes through in the course of the year. It has also been known since antiquity that the vernal equinox very slowly moves in the opposite direction around the zodiac, completing a revolution in about 26,000 years, called the Platonic Year.

In giving names to the twelve constellations of the zodiac, early men expressed in imaginative terms their direct experience of the twelve images as characters of a script recording the deeds of the sun. In other words, early men did not regard the movements of sun, moon and stars in relation to the earth as mechanical and predictable facts, but as manifestations and records of divine, mythological activity.

Just as the zodiac can be and was used as a measure to gauge actual as well as divine activity in the universe, so also was Adam Kadmon, the archetypal man, created in the image of God and standing within the twelvefold round of the zodiac experienced as the measure of mankind and the universe.

While in the pre-scientific era of the history of mankind, the far-reaching influence of the sun and the planets on the earth and on man was unquestioned, the materialistic-mechanistic view of science in the nineteenth and early twentieth centuries found this difficult to accept. Even the influence of the moon on telluric phenomena was disputed and largely regarded as superstition, albeit today, the interaction between moon and earth manifesting in tides and in the movements of water in general, as well as in plant-growth, is well-established.

That the planets, too, might have an influence on telluric phenomena is still widely discounted, although during the last few decades, interactions between the various bodies of the solar system have been postulated. M. Gauquelin (1973) has collected a mass of challenging observations which offer formidable evidence that highly relevant interactions exist between planetary movements and biological phenomena.

Our concern here, however, is not with causative influences or dependencies, but rather with aspects of meaning and significance. In this context, the following may be of interest. I have mentioned the 26,000 year rhythm of the vernal point of the sun which, from antiquity on, has been called the Platonic Year. We find this rhythm again in the microcosm of human existence, for when a man is at rest, he breathes on average about eighteen times a minute which amounts to 26,000 times in a twenty-four-hour day; and in an average life span of just over seventy years, there are 26,000 days which would be one cosmic day within a 'cosmic' year of 26,000 years.

Having tentatively suggested that man as a microcosm could be seen as a resonance of the macrocosm, I shall return to the statement with which I closed Chapter 7 — that the cosmic, zodiacal form of man, having given rise to the threefold embryonic disc, will call forth the human form in the course of the forty days of the second phase of embryonic development.

First of all, let us take up the question as to how the spherical, twelvefold zodiac can meaningfully be imagined as having a bearing on the three-layered germ disc with its three spatial axes within the fourfold sheaths of embryonic development, consisting of chorion, allantois, amnion and yolk-sac, and what its part may be in the creating of man 'in the likeness of God'.

According to tradition, the circle of the zodiac consists of four triangles: one for fire, one for air, one for water, and one for earth, but also of three squares, each containing at its corners one of the four elements.

This three-times-four or four-times-three structure, while representing the ultimate boundary of the solar system, is at the same time spatial and finite. If we are to experience it as giving rise to a form 'in the likeness of God', we have to include the infinite as an attribute of God, relevant to his nature and likeness.

While we can mathematically think the infinite as any number divided by zero, and hence grasp that there is only one zero and one infinity, it is not so easy to imagine this visually and geometrically. We know that two straight lines in one plane have only one point in common and that two parallel lines intersect at one point only at an infinitely distant point. We also know that this intersection takes place in both directions of the lines, to the left and to the right, yet in the same infinity.

We may be tempted to imagine space as spherical, curved and closed in itself and a straight line as ultimately circular and returning to itself. While this would provide an adequate image for an unbounded quality of space, it would not include the quality of infinitude which remains essentially open ended. If we could train ourselves to imagine infinitude as singular yet unlocatable, unbounded and open, yet comprising all the elements of form, point, line and plane, then we could imagine our likeness to God reduced to the simplest level.

Let us now return to the question of how the zodiac can be meaningfully related to the human form as it develops in the second embryonic phase.

The word 'zodiac' meaning 'circle of animals' may arouse associations with evolution in that consecutive stages of the development of the embryo during the forty days of the second phase could be seen in terms of successive evolutionary forms of the animals of the zodiac.

Only seven, however, of the twelve signs are images of animals (Figure 15); three are human, one is half-man, half-animal, and the last, Libra, or Scales, is a mechanical device. On the basis of the theory of evolution, moreover, one would expect some order or progression in the sequence of the animal pictures, but no trace of this can be found. Out of the five vertebrates represented in the circle, three are horned ungulates of which two, Ram (Aries) and Bull (Taurus) are placed side by side; the third, Capricorn (Goat) is separated from them by Fishes (Pisces) and Waterman (Aquarius). The only other mammal, the carnivore, Lion stands opposite Waterman between one of the two arthropods, Cancer (Crab) and the human Virgin, and the other arthropod, Scorpion which is opposite Bull, stands between Scales and Centaur (Sagittarius).

As has been said, the position of the seven animal forms in the zodiac does not seem to indicate any evolutionary order or progression. In connection with the discoveries in modern biology described in the second part of this book, we would in any case expect the principle of evolution to be relevant rather to the genetic than to the form aspect of development, rather to differentiating detailed features than to the overall form or gestalt.

Medieval alchemists, however, saw the upright human form in terms of the twelve signs of the zodiac (Figure 13); Ram was

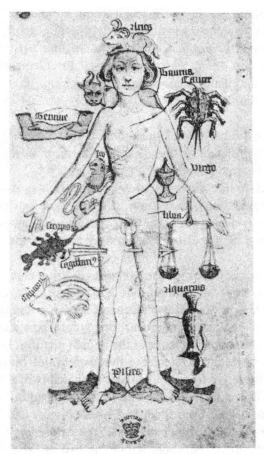

Figure 13. Zodiacal man. A medieval depiction of the relation of the zodiac to the body.

related to the head, Bull to the throat, Twins to the clavicles and shoulders, Crab to the chest, Lion to the heart, Virgin to the belly, Scales to the hips, Scorpion to the genitals, Archer to the upper arms and thighs, Mountain Goat to the elbows and knees, Waterman to the lower arms and calves, and Fishes to the hands and feet. This is the same correlation ascribed to the ancient Adam Kadmon, the spherical man expanded into the universe (Figure 14).

Medieval alchemical illustrations also frequently depict correlations between the planets and the seven life organs of the body

8. EMBRYONIC FORM DEVELOPMENT AND THE COSMOS

Figure 14. Indian depiction of Adam Kadmon.

— Saturn, Jupiter and Mars being seen in relation to spleen, liver and gall bladder; sun to the heart; likewise Mercury, Venus and moon to lung, kidney and brain respectively, and moon to the gonads as well, but a contemplation of these alchemical concepts would be outside our actual subject.

We may be aware that the vernal point of the Sun has moved from the constellation of Ram to the Fishes and that we are seeking for a link between the zodiac and the becoming of form in embryogeny. We should also like to recall that earlier on we argued that the significance of the zodiacal signs lies in their making visual and measurable the changing interrelationships between earth, sun and the planets. In following this up, an approach developed by F. H. Julius may be helpful. While adhering to the traditional signs of the zodiac, Julius analysed the yearly, seasonal changes in the relation between sun and earth according to the position of the ecliptic (in summer mainly above, in winter mainly below the horizon) and to the upward and down-

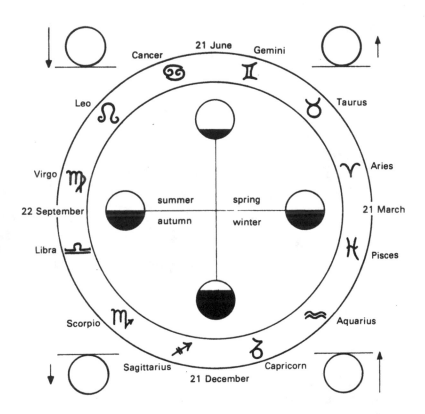

Figure 15. Annual movement of the sun through the signs of the zodiac.

ward movement and the acceleration or deceleration of the movement of the ecliptic. (Figure 15).

Julius described the changing interrelationship between sun and earth mainly from the aspect of the polarities of light and darkness, levity and gravity, and from them developed dynamic images of the zodiacal signs. We could assign these characteristics to the corresponding constellations. Thus we can see Mountain Goat, with its powerful upward thrust of butting horns and continual climbing of rocky heights, as the force which wrests itself away from the pounding hoofs of Centaur (Sagittarius), and frees the accelerating upward-rising power of light, which mounts with increasing speed. Waterman and Fishes hold the transition between winter and spring, from darkness to light; a balance is reached in the equinoxes when the acceleration in Fishes, becomes a beginning deceleration in the Ram, or when the darkness recedes before the light which becomes victorious.

The Ram with his head turned back and his horns swinging round in a spiral movement, stands after the equinox in the increasing light, but is decreasing in upward impetus which then almost comes to rest in the powerful inertia of the Bull, finally stopping in the warm playfulness of Twins at midsummer.

Following Twins stands Crab. Here the change to downward movement is expressed in the tendency to inversion, retreat, to embrace, rather than the backward-looking tendency in Ram. The golden days of late summer and the harvest are reflected in Lion and Virgin.

Scales stands at the entry into darkness between the fertile and receptive Virgin and the aggressive sting of Scorpio who turns death against himself plunging into darkness, and an ultimate standstill in the depths of winter is reached in Centaur before the upward dynamic in Mountain Goat can begin anew. Capricorn and Ram seem to possess a somewhat outgoing yet divisive character, while Crab and Scales seem to have an embracing, enclosing, uniting gesture.

These are, of course, very rudimentary attempts at describing formative qualities in terms of the twelve signs of the zodiac, but they are qualities and dynamics which can be discerned in embryonic development, and for this reason we bring them to the notice of readers of this book.

With these deliberations, we have taken up what was stated at the end of the previous chapter. The expanded spiritual human gestalt which unites in the moment of conception with the spirit germ in the ovum develops in the second embryological phase into the actual human form. Here the zodiacal forces are from the beginning generally valid human form principles. That these principles which are manifest in every human form are subject to a wide variety of differentiation and individualization, partly determined by genetic factors, is a matter of course. We should also not think that the developing and ultimately completed human form is merely a transcript or copy of the zodiacal image, which I hope I have made clear. It is a question of form principles which are similar to those demonstrated by the so-called Chladny sound figures, which initiate the specific direction of development in the becoming human gestalt.

We shall now turn to the second phase of embryonic development, the embryonic phase proper. At the beginning of Chapter 7, we described the six stages of the first phase of development in which the three-layered germ disc develops from the fertilized ovum. We mentioned that from each of these layers, one of the three primary organ-systems of the body will develop: from the ectoderm, the nervous system and the skin; from the mesoderm, blood, bone and muscle that give form and movement to the body; and from the endoderm the digestive system and the great inner organs of the body.

In the first half of this century as also in the nineteenth century, it was assumed, largely as a result of observations of animal development, that ectoderm and endoderm developed first and only then gave rise to the mesoderm as a secondary event. More recently, several authors, notably König and Wilmar, have suggested that ectoderm and endoderm may well be secondary, evolving out of an earlier, pre-existing and undifferentiated mesoderm which surrounds the embryonic disc as extra-embryonic, peripheric mesenchyme.

Reading the descriptions and comparing the photographs and drawings in relevant publications, one gains the impression that we are not confronted with an either-or situation, but rather with an indeterminate process of threefoldness as it so fascinatingly appears in Chinese philosophy in which the Tao is both the secondary harmonization of the polarities of Yin and Yang as well as their primary origin.

A dynamic concept of a trinity as indeterminate processuality has been richly developed by Jean Gebser in his book *Verfall and Teilhabe*. The mystery of the Christian Triune probably also includes this element of transcendence. The quality of indeterminateness, of openness, which we encounter, too, in modern quantum physics appears as a fundamental element in embryogeny, if we only recall the search for the 'organizer' conducted by men like Spemann, Goodwin and Sheldrake.

Through test tube fertilization, it has recently become possible to observe the early stages of the first phase of embryogenesis as it actually unfolds before one's eyes, but our knowledge of the second phase is still dependent on, and limited to, isolated instances of embryos in early stages of development having been accidentally discovered.

The first impression one gains from the study of successive stages of foetal development described in various textbooks is one of puzzled bewilderment. But when one tries to follow with one's imagination the astonishing transitions from form to form, one begins to experience a creative power of dynamic movement. This movement seems to consist of both cell-division and multiplication, of growth as well as of actual migrations and displacements of cells. Our concern here is with the *form* to which these dynamic processes lead and with which the characteristics of the zodiacal constellations could be imaginatively linked.

Embryonic form-development, however, is so complicated, the dynamic processes so intertwined and the material so vast that only some rather tentative and isolated suggestions can be made here. Therefore, a very limited, partial and aphoristic interpretation of embryogenesis along these lines will be attempted.

Towards the end of Chapter 7 we described the final, sixth stage of the first phase of embryonic development, which is at the same time the starting-point of the development in the second phase. The originally round embryonic disc with its three germ-layers has become elongated, differentiating a broader, cranial, or head-end, from a narrower caudal, or tail-end, thereby establishing the three-dimensional orientation of the future human body.

The first differentiating movement that heralds the embryonic phase proper is the appearance of the chorda which grows as a hollow, mesodermal invagination or tube from the middle of the ectodermal plate forward between ectoderm and endoderm. In its spearlike thrust, it marks out the direction of man's ultimate upright posture which will be vested in his spine and skeleton. But to begin with, the central, midline, upward thrust of the chorda seems to call forth a threefold, paired response in the

111

three germinal layers. In the ectoderm, the floor of the amniotic cavity forming the dorsum or back of the embryonic disc, two parallel ridges appear growing from the middle cranially (headwards) and caudally (tailwards). These two ridges will fuse later in these two directions to form the neural tube, the beginning of the central nervous system of brain and spinal cord.

In the mesodermal layer, the appearance of the chorda is followed by the bilateral, paired formation of the first somites, cubelike formations on either side of the chorda, which will multiply to forty-four pairs and will ultimately give rise to the bony skeleton and all the muscles.

The endodermal response to the chorda-thrust and the connected lengthening of the originally circular embryonic disc is the initial appearance of the foregut cranially and hindgut caudally from the originally wide-open communication of the yolk-sac. This twofold gut-development is the beginning of the formation of the future intestine from which the various inner organs such as lung and liver will develop.

Bringing our power of imaginative thinking to bear upon these, after all dramatic processes, we could see in the dynamic upward and forward thrust of the chorda, calling forth a threefold pairing, something of the zodiacal picture of the Archer, the half-man, half-horse, aiming his arrow towards the Twins on the opposite side of the zodiac. Chiron, the Centaur (Sagittarius), was the renowned primeval teacher of the first generation of the Greek heroes such as Heracles, Orpheus and others, fathers of human existence in the becoming. In early embryogenesis, the Centaur or Archer seems to fulfil a similar function.

Let us now regard the first developmental steps in each of the three germ-layers separately. We have stated that on the surface of the ectoderm which is the upper or dorsal layer of the embryonic disc (but the bottom layer of the large and growing amniotic cavity), there appear two parallel longitudinal ridges in the cranial-caudal directions. They grow the whole length of the dorsal side of the ectodermal plate, forming the neural groove which, to begin with, is open to and communicates with the amniotic fluid above. The leading edges of the two ridges which are called the neural crest first grow outwards and upwards, in cross-section like two outstretched and raised arms, then closing in towards one another as though in an embrace, thus forming the neural tube which is filled with amniotic water. The fusion begins in the middle of the neural groove and proceeds towards both the cranial and the caudal ends, until complete closure is accomplished. The closed neural tube then sinks below the dorsal surface of the ectoderm which later becomes the overall cover of the skin. The neural crests which had brought about the fusion first form an intermediate zone between the neural tube and the surface ectoderm, but then divide again into the paired spinal ganglia on either side of the neural tube.

All this is completed within the first few days of the second embryonic phase. We may discern in this sequence of form-processes, three formative principles: there is the arrow-thrust of the Archer wresting uprightness out of spherical non-direction; and the other principle of doubling or twinning in which we may see the characteristics of Twins, often depicted as two separate children holding each other's hands, signifying the left-right bilateral symmetry, yet striving apart. Lastly, there is the third formative principle in the movement of fusion, of embrace and closure which can be seen in the gesture of the Crab, or in the uniting, balancing gesture of Scales.

The gesture of closing, healing-over of the ectoderm forming the outer skin, may be linked to the descending quality of the sign of the Virgin, which leads the descent of the light into the darkness.

The neural tube now lying below the ectodermal skin-covering, flanked by the mesodermal development of the somites and above the endoderm, grows in thickness, but particularly and much more, in length. The cranial end of the tube begins to grow downwards as does the caudal or tail-end, forming an arch which recalls the body of the goddess Nut arching over the earth as the heaven (Chapter 1) in the mythological Egyptian concept of creation.

The growth in length seems to exceed the overall growth of the embryonic disc and brings about multiple doubling-up in deep folds in the cranial and middle part of the neural tube, forming what will become the central part of the brain. The cranial end of the neural tube now develops a bilateral bulge which first grows outwards and then downwards, forwards, upwards, backwards, downwards and forwards again, ultimately completely covering the central part of the brain as the two cerebral hemispheres. The form of the Ram's horn here becomes obvious both in the growth-dynamic as well as in the outer form and in the form of the inner hollow or side-ventricles which arises as the central cavity of the neural tube.

To continue these deliberations, we may see in the powerful increase in length in this stage of embryological development an element of Capricorn, rising with growing speed from the darkness into the light, and in the multiple bending and flexing of the neural tube, the form-element of Waterman approaching a balance between darkness and light; then the branching into two separate, opposite bulges in the pairing and polarizing characteristics of Fishes, and finally the wonderful development of the two hemi-spheres of the brain in Ram.

By this time, the twelve pairs of cranial nerves and the thirty-one pairs of spinal nerves have begun to grow out of the developing central nervous system and the sense organs have begun to form. Of these latter, we shall briefly describe only the development of the eye, as an example.

From the lower frontal part of the brain bilaterally, a bud grows out towards the overlying skin. In the corresponding part of the skin, a blister of cells is formed which begins to grow inwards, gradually becoming the lens of the eye, while the outward-growing bud from the brain forms a cup, becoming the retina and the optic nerve. The outward-growing cup receiving into it the lens may recall the relationship between the fibula and the patella. Knee and elbow are seen traditionally in the sign of Capricorn which marks the ascent from darkness into light and are in terms of the bony structure of the body correlations of the organ of sight.

When we consider the fact that both lens and cornea, originally living cellular tissues permeated with blood, ultimately become completely transparent and nearly dead structures (the cornea of a dead person can be transplanted into a living one), we may think of the death-sting of Scorpio who leads the descent into darkness, yet also into the resurrection, into new light in handing over, as it were, to the next constellation, the Archer. The sinking down of the ultimately transparent lens into the dark embrace of the cup of the retina, on the other hand, has something of the gesture of Crab. In all this though, we must not forget that retina, lens and cornea are only some of the essential ingredients of the eye and it would seem probable that in the formation of any one organ, the formative gestures of many or all the zodiacal elements may be involved.

We shall now return to the development of the endodermal germ-layer. Towards the end of the first phase of embryonic development, this layer is circular, the roof of the relatively large yolk-sac below the relatively smaller amnion. With the powerful longitudinal growth of the germ disc, the arching neural tube and skin of the ectoderm as well as the growing mesoderm, the so-called head and tail folds are formed cranially and caudally. The roof and upper part of the yolk-sac, which we may liken to ancient Gaia, the earth spread out below the arching neural tube of Uranos, the encircling heaven of early Greek theogeny, have by necessity to follow this growth development, but the rest of the yolk-sac invaginates as foregut and hindgut respectively into the head and tail fold, becoming embraced both cranially as well as caudally, but also to the left and right, by the ectodermal and mesodermal growth movements.

The forward and backward push of the invaginating foregut and hindgut can be seen as the powerful yet retarding thrust of the Bull before the sun's course enters the bilateral embrace of Twins, and the head-tail embrace of Crab, of both mesoderm and ectoderm. From the growing foregut, a number of further protuberances emerge, giving rise to liver, gall bladder, pancreas and stomach as well as trachea and lungs. These gut invaginations develop into the complete intestinal and digestive tract. The hindgut which has taken the allantois with it in its invagination, first develops a common cloaca and then divides into rectum and bladder.

I should still like to draw attention to one of the most striking polarities between the development of the ectodermal neural tube and the endodermal gut. The neural tube starts as an open tube, fusing from its middle part both towards head-and-tail ends, enclosing some of the amniotic fluid above, but becoming a completely enclosed inner space. The gut, on the other hand, is a closed double cul-de-sac to begin with; it grows into many secondary, always closed protrusions, but finally has to open up both cranially as well as at the tail end in nostrils and mouth, anus and urethra. The intestine thus becomes an open thoroughfare for substances to enter and to leave the body. Here we may perceive a quality of Scorpio, a turning-point from darkness to light, from inside to outside.

We come now to the development of the third — or also the primary — mesodermal germ-layer which both separates the other two layers of ectoderm and endoderm, but equally unites them in an all-embracing periphery. In this light, we may see the mesoderm as Eros, the third principle which arises between Uranos, the heaven, and Gaia, the earth; or also as the first principle in the Egyptian mythical system where Atum, the whole one, brings forth without female partner the double nature of the atmosphere, Shu and Tefnut, and separates the ultimate polarities of Nut, arching as heaven above, from Gebeb, the earth below, yet as Atum containing and embracing all. In this connection, it is interesting to note that Rudolf Steiner described the double nature of the mesoderm as the origin of ectoderm and endoderm (1983).

Before going into the actual mesodermal form development, it may be helpful to describe the changing position of the embryo in its sheaths within the womb. During the first phase of embryonic development from nidation onwards, the contact between trophoblast and the wall of the womb, the beginning placenta, was over the segment where the amniotic cavity had developed. Towards the end of the first phase and increasingly during the second phase of embryonic development, the rapid and mainly cranially directed growth of the embryonic disc and amnion moves away from the broad contact with the developing placenta, bearing the relatively diminishing body-stalk at the caudal end as the link to the placenta. This body-stalk will later be called the navel cord when the growing embryo has turned by nearly 180° to its normal, head-down position within the womb and its sheaths, the insertion of the navel cord having moved from the caudal to the ventral side of the growing embryo.

Already by the end of the first phase, soon after the emergence of the blood-islands and immediately following the growth of the chorda, there appears in mesodermal tissue and cranially outside the embryonic disc, the small pericardial cavity, the outer layer of the heart. Into this heart, the blood vessels which had resulted from the confluence of blood-islands integrate, to become step by step, the four-chambered heart in everchanging and altering movements and form processes. It is important to hold on to the observation that the starting-point of this development is the independent, primary rhythmic movement of the blood-islands that integrate into an especially prepared space that finally brings about the form of the heart.

These form processes are so manifold and complicated that I shall not attempt to describe them. It is, however, interesting to note that the traditional zodiac sign for the the Lion, ♌, seems to be a simplified symbolic abbreviation of the first gesture of the form-development of the heart. We should not forget the unique phenomenon that the heart is the only organ that does not develop within the embryonic disc, but outside it, in front and ahead of it, becoming enclosed in the body of the embryo only as it gradually grows forward and over the developing heart. Even in embryos of five and six weeks of age, the heart still seems like a huge bulge only just held within the tiny embryo. Here again, we witness the twofold or ambiguous creation from within of the rhythmically pulsating blood-islands, on the one hand, and the peripheric, embracing, interiorizing growth of the body of the embryo.

While the heart as a singular, central organ, unites the largely bilateral vascular system, there begins to develop a paired meso-dermal organ within the middle part of the embryonic disc — the beginnings of the kidneys which, through many transformations descending tailwards, will link to the allantois which, at the beginning of the development of the heart, lies in the body-stalk as far out caudally from the embryonic disc as the heart does cranially. From this latter development, the urogenital system will gradually arise in the fifth to seventh week.

I mention this because it is at this point that a discrepancy between the biblical Genesis and embryogenesis appears. Moses describes the transition from androgynous man to bisexual man as starting with the male, the female being secondary. Embryonic evidence, however, shows the female sexual form to be nearer the earlier androgynous form of man, as far as one can ascertain it, the male form arising as a further step. (In both sexes, the gonads are originally sited in the abdomen where they remain in the woman, but from where they descend into the scrotum in the man.) Perhaps we may see in both the cardiovascular system as well as in the urogenital system, Eros as the middle element which brings interaction and creativity between Uranos, the heaven (the nervous system of the ectoderm), and Gaia, the earth (the polar system of the endoderm).

The mesoderm, however, does not only bring about these two systems, but it is also responsible for the entire bony structure of the body and all the muscles, thus supplying the form as well as the movement of the human body. Could this element be experienced as the Egyptian Atum, the whole one, the mesenchym or mesophyl of which Wilmar and König speak?

We shall now briefly describe the gradual appearance of the human form, the gestalt, which is completed within the forty days that follow the first twenty-day phase of development, thus ending about the sixtieth day after fertilization of the ovum. During the first ten days, the embryonic disc develops into a somewhat elongated, round structure, strongly curved, reminding one perhaps of a fat caterpillar. The ventral side of this early form shows four main parts: the head-end, bending forwards and protruding above and beyond the big pericardial swelling of the heart which has just become part of the body of the embryo. Below the protruding bulge of the heart, the umbilical cord emerges, the result of the body-stalk having remained behind in growth, when the amniotic cavity with the embryonic disc grew mainly forward and downward; a direction which seemed indicated by the zodiacal gesture of the Bull that pushes forwards, and Lion which claws the sun downwards into increasing darkness.

The movement of enclosing the belly which earlier was open

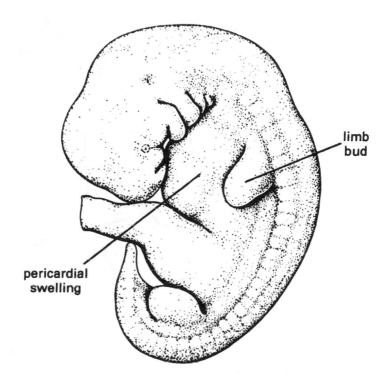

Figure 16. Human embryo at five weeks (7 mm).

towards the yolk-sac can be related to the quality of Virgin, and the fourth part is taken up by the tailfold which, relatively large at the time, will then remain behind in growth so that by the end of the embryonic phase, it will have disappeared. We may see this curling back on itself and subsequent disappearance as a Scorpion element.

By the fifth week (Figure 16) the whole structure measures only seven millimetres. At this time, the head looks rather like a whole body with a head, trunk and limbs. What seems to be the head is really the developing forehead; the face, however, has not yet begun to fuse from its paired beginnings which will later form the nose, mouth and throat. Meanwhile, two of these paired processes — that which will form the lower jaw, and another which will. participate in the formation of the throat — are so prominent that

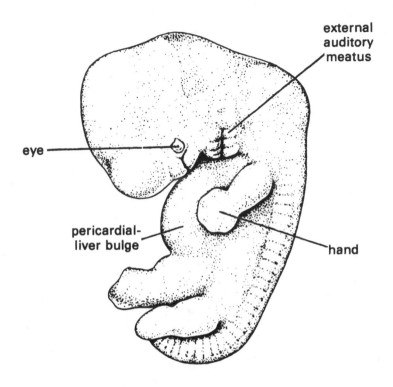

eye

external
auditory
meatus

pericardial-
liver bulge

hand

Figure 17. Human embryo at six weeks (13 mm).

they look like upper and lower limbs. It is only during the next
ten days when the actual limb-buds appear and the face begins to
take shape that one realizes that what had looked like a whole
body is indeed only the head. It would seem that the head were
starting as a body, becoming pure head only in a secondary transformation of the body form when the body proper has begun to
develop. And yet already by the fifth week, the head part shows
both the optic placode on both sides of the forehead and the otic
— or ear — placode more towards the back of the head. From
the optic placode the eye will develop, and from the otic vesicle
the inner ear. With regard to mere form development, however,
we are justified in seeing all elements of the whole body-form
appearing already in the early head development.

In the course of the next week (Figure 17), there appears the upper limb-bud on either side of the pericardial swelling, then the lower one on either side of and below the umbilical cord. To begin with, both are unshaped, undifferentiated rounded buds, the upper one just beginning to indicate hand and forearm. The main growth and development has taken place in the head which is strongly flexed forward, lying with its slowly forming face over the pericardial swelling within which the four-chambered heart can be discerned as well as the liver.

Through the transparent skin of the head, the deep flexure of the growing neural tube can be seen on the top, with the first indication of the forebrain vesicles from which the two hemispheres will develop in the region of the forehead.

The most important development is that which takes place in the region of the face. The mandibular processes have fused medially, the maxillary processes arc growing below the bulky development of the nasal folds. Yet even around the sixth week when the embryo has reached a length of 13 mm, one would not necessarily recognize it as a potential human form.

Particularly the characteristic gesture of potential uprightness, the development of detail in the limbs and the 'emancipation' of the head from the rest of the body, important and crucial factors which constitute the signature of the human form, are not yet manifest.

By the seventh week, the embryo has reached a length of 17 to 18 mm (Figure 18). By now the upper limb-buds have differentiated into hands with fingers, into lower and upper arms, although the shoulders are not yet obvious. The lower limb-buds show just the beginnings of toes; the thighs are not yet clearly discernible.

To conclude our contemplations on zodiacal relationships in embryogenesis, we can see from the progression just described that the limbs develop formwise from distal to proximal, from hand to shoulder, and from foot to hip, corresponding to the constellations of Fishes to Waterman, Capricorn and Archer to Scales in the hips and Twins in the shoulders.

In the face, the maxillary processes have fused, forming the mouth with the mandibles. Also the nasal folds have fused and the medial nasal processes have grown to form the nose. The outer ear is still very rudimentary and rather low in its position

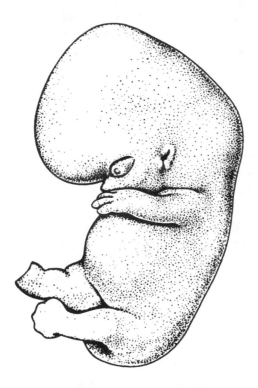

Figure 18. Human embryo at seven weeks (18 mm).

behind the mouth, while the eyes are now recognizable although lacking eyelids. Head and brain have grown considerably. The heart and liver bulge is still very prominent, but the tailfold has become quite small.

Around the eighth week (Figure 19), the brain hemispheres have grown to cover about half the brainstem, giving the head a much rounder appearance. The eyelids have begun to shape, the face is now recognizable. The outer ears have gained their ultimate position, but are still rather primitive in form. The head is set off from the body by a short neck, but there is no back curvature as yet. Fingers and toes, arms and legs as well as shoulders are clearly apparent and the sex of the embryo can be perceived. The miracle of the human form has been achieved.

I hope I have been able to draw attention to a dimension in

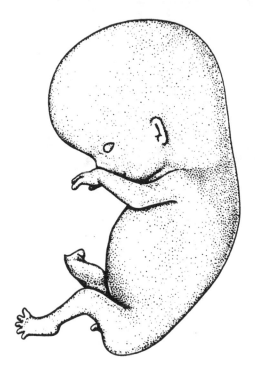

Figure 19. Human embryo at eight weeks (30 mm).

human becoming in which macrocosmic form principles emerge into appearance in the microcosmic human form. In the following chapter, I shall refer to certain spiritual-historical realities which can demonstrate the relationship of the macrocosmic to the microcosmic.

Biology and Spirit

9. Embryogenesis and the Gospels

Aristotle's and Thomas Aquinas' idea of a vegetative, sentient and rational soul reappeared at the beginning of this century in the work of Rudolf Steiner, who took up the ancient concept of reincarnation in a new and contemporary way. In many publications and lecture cycles, Rudolf Steiner gave very detailed descriptions of the experiences the human individuality goes through between death and a new birth. While Steiner speaks of bodies rather than souls, his approach is entirely spiritual and imaginative. Death is described as a gradual, step-by-step separation of the four elements that constitute human existence on earth. First the etheric or life body which corresponds to the vegetative soul separates from the physical body and begins to loosen and free itself. This calls forth the so-called 'tableau', which is a detailed panorama of the past life, as it has severally been described in recent books about after-death experiences of people who have been resuscitated from clinical death.

Then the individuality begins to expand through the planetary spheres. The astral body, which corresponds to Aristotle's and Aquinas' sentient soul, also begins to widen and to dissolve, while the individuality, the actual ego or rational soul according to Aristotle, expands into the ultimate periphery of the cosmos, the zodiac.

Rudolf Steiner describes this journey in astounding detail, as well as in exceptional beauty, and states that the individuality, when totally expanded during what he calls the 'midnight hour', hears a summons to return to the earth and sets out on a descent, again through all the planetary spheres, encountering, learning from, working with a manifoldness of hierarchic beings. Helped and supported by these beings, man develops during this descent the spiritual-physical form or 'plan' of his earthly body. Steiner calls this form the 'spirit-germ'.

On the further descent through the planetary spheres, the individuality again surrounds itself with the sentient or astral element and later with the vegetative-etheric body, finally awaiting its new earthly incarnation while still in the sphere of the moon.

In the moment when conception takes place the individuality, now clad in astral and ether bodies, lets its spirit-germ sink down into the fertilized ovum. The three spirit-soul elements of ego, astral and ethereal delay their descent into the embryo for seventeen to twenty days and then only connect themselves with the warmth, air and fluid integuments within which the embryonic disc has developed.

We are making this brief and rather inadequate reference to Steiner's descriptions mainly for the sake of his concept of the spirit-germ. Not only is his reference to the time around the seventeenth to twentieth day remarkable, as this coincides with the end of the first phase of embryological development, but more important is the idea of the physical form of the human body originating in the expanse of the cosmos, independent of heredity.

I wish to relate here a personal experience. When, at the beginning of the Second World War, I was interned as an enemy alien, my wife was expecting our first child at home. I was at the time preoccupied with the intense study of some of Rudolf Steiner's work, but was also anxiously awaiting news of my child's birth. I repeatedly experienced silvery white, vortex-like forms descending over the night sky. In their appearance, they reminded me of the form of the human uterus with the Fallopian tubes, also of the larynx and pharynx with the Eustachian tubes, and the human breastbone with the clavicles winging out to both sides. I came to the conclusion that these shining forms were images of descending spirit-germs.

The three organic structures I have just referred to have an obvious similarity in form (Figure 20). The uterus would not only be the place within which the embryo and foetus develop, but according to Democritos, it plays an essential part in the development of the *form* of the embryo. More recently, E. Blechschmidt developed this theory in great detail and we mentioned in Chapter

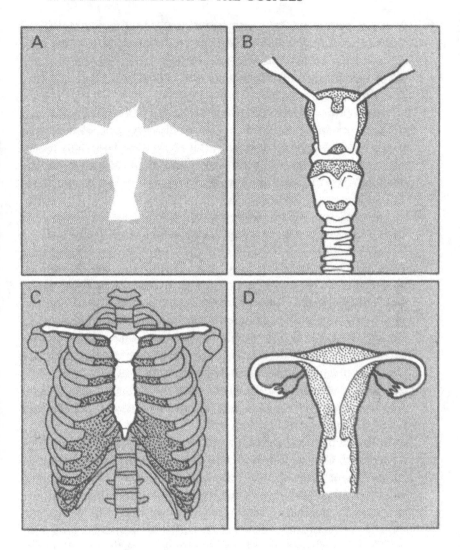

Figure 20. The dove form in larynx, oral cavity, and Eustachian tubes (B); in breastbone and collar-bones (C); in uterus and Fallopian tubes (D).

6 Edwards's geometric transformations between the forms of vortex and uterus, which had not yet been worked out at the time of my experience. There is also an old occult prophecy which holds that in a far distant future, man will create his progeny by speaking through his larynx.

The breastbone with the collar bones has often been acclaimed as the signature of the form of man. Ultimately, I see in the image of the spirit-germ the likeness to the white dove, the Holy Spirit, that in so many paintings is seen descending upon Mary in the Annunciation, upon Christ in the Baptism in the Jordan, and again upon Mary at Pentecost.

The essential importance, however, of Steiner's description of the spirit-germ lies in the fact that it established a duality in the event of conception — a cosmic conception of form as distinct from the physical, hereditary one, thus confirming the fundamental discovery in biology of this century, relating to the distinction between the strictly pre-determined genetic aspect and the epigenetic aspect of the development of form. This discovery can only be understood in its far-reaching human relevance, when we realize that the genetic determines only what distinguishes one man from another, that is the differentiating details of each man's make-up; but that which makes every man recognizably human, that which we all have in common is not genetically founded, but originates elsewhere.

This well-documented realization on the part of modern biology allows a new interpretation and understanding of Christianity. So far, theologians (possibly for good reasons) have vehemently resisted any tendency to regard the Gospels other than as historic documents. This postulated historicity has meant in the mental climate of past centuries that the central and decisive events in the Gospels such as Mary's conception and particularly the Resurrection were regarded as miracles, which means as facts rationally inexplicable, because they are outside or contrary to natural laws.

Later on, we shall argue the falsity and destructiveness of the concept of natural laws. Here we want to suggest that on the basis of contemporary insights in biology, we could more deeply experience and understand the central events in the Gospels as mythos rather than as miracles, which means as revelations of timeless archetypes in humanly moving images.

The New Testament begins with the Gospel of Matthew which lists the genealogy of Jesus, starting with Abraham, going through David and his son, Solomon, ultimately to Joseph, the husband of Mary who gave birth to Jesus. This impressive line of ancestors which opens the Gospel of Matthew like a trumpet call, is followed by the account that Mary found that she was with child by the Holy Spirit which seemed to both parents to precede their marriage. It is further explained to Joseph by the angel that it is by the Holy Spirit that Mary has conceived this child. Yet all this does not necessarily gainsay the validity of the genealogy at the beginning of Matthew's Gospel; it can rather be experienced as a fulfilment of the prophecy that 'a virgin will conceive a child'.

In conformity with general custom, the Gospels give only the father's genealogy, thus attaching the genetic element to the male. While this is probably largely due to the patriarchal bias of Judaic Christianity, the attributing of the genetic element to the male and the element of form to the female finds interesting support in the polarization of male and female germ cells into sperm and ovum, the sperm being composed almost entirely of nuclear substance which carries the genes, while the ovum, though obviously having a nucleus, is mainly cytoplasm, which is thought by some authors like C. H. Waddington, M. Fishberg and A. W. Blackler to be more involved in the element of form.

The Gospel of Luke, on the other hand, opens with a description of the annunciating angel's visit to Mary who is betrothed to Joseph, and speaks of the conception of Mary's child through the Holy Spirit. Only after the birth of the child Jesus, are Joseph and Mary referred to as his parents, and only following the Baptism of Jesus in the Jordan, described in the third chapter of the Gospel of Luke, when the Holy Spirit again descends like a dove, now upon Jesus himself, does this Gospel list his genealogy, which this time starts with Jesus and Joseph and leads back through Nathan, another son of David, to Abraham, Adam and ultimately to God.

These genealogies have been taken to express 'what people thought', but what they really reveal is a knowledge of the two conceptions — the genetic, and the separate conception of form through the Holy Spirit, the spirit-germ, the White Dove.

The essential message in both Gospel accounts is that the conception of the human form in its own spiritual origin was for the first time revealed here, while giving full value to the importance of the genealogy and its genetic relevance. It is a timeless insight, holding good for all human conception. It is not specific to the conception of Jesus and yet it had never before been perceived or declared.

Even now, nearly two thousand years later, this message has still not been recognized and accepted, although the Annunciation or Conception story is probably one of the most widely known and has been illustrated by countless artists throughout the centuries. By having made it into a miracle pertaining exclusively to the conception of Jesus, men have upheld 'what people thought' in relation to our own conception. The continuing conviction that we have a hereditary origin only, that we are the final product of animal evolution, that we are ultimately nothing other than our 'selfish genes', has deeply moulded our self-evaluation; it has prevented us from facing the fact that we are meant to go the way from creature to creator.

The accounts of the twofold conception of Jesus Christ are particularly relevant for an understanding of the Resurrection of Christ. John, the only one of the Gospel writers present at the Crucifixion of Christ, gives a detailed account of the events of the early Sunday morning after Good Friday. He reports that he and Peter were told by Mary Magdalene, who had visited the tomb in which the body of Christ had been laid while it was still dark, that she had found it empty. Peter and John both run to the tomb and likewise find it empty. The two disciples leave the tomb again, but Mary Magdalene remains. Weeping, she peers into the inside and sees two angels in white and speaks with them. Then turning around, she sees a man standing without, whom she does not recognize. He asks her: 'Why are you weeping? Who is it you are looking for?' Thinking it is the gardener, she says: 'If it is you, sir, who removed him, tell me where you have laid him, and I will take him away.'

How is it possible that Mary Magdalene, who was so intimately connected with Jesus during the last year of his life, who had stood beneath the cross when he was crucified, did not recognize him in the Risen Christ? Only when he calls her by name does she suddenly know who it is.

We encounter a similar episode when the disciples go fishing in the sea of Tiberias. As they return to the shore, they see Jesus standing there by a fire of coals, but do not recognize him. Only after he speaks to them does John say: 'It is the Lord' — but as the disciples step ashore, none of them dares yet to ask: 'Who are you?' (John 21:1–12).

In the Gospel of Luke, there is a moving account of two disciples who, walking home to the little village of Emmaus on the evening of the same Sunday, meet Jesus on the way. They take him for a stranger and converse with him. When they reach their home, they invite him into the house and only when he blesses and breaks bread with them, do they recognize him.

All this indicates that the Risen Christ is not the same as the one who died on the cross. This is not a resurrection like that of Lazarus or the young man of Nain or the little daughter of Jairus all of whom had 'died' and were brought back to life by the power of Christ.

What then is the nature of the body that Christ raises in the Resurrection? It is the non-material human form without distinguishing genetic features; universally human but with no individualizing characteristics. The Resurrection Body reveals two completely new phenomena: the one is that form as such becomes a visible experience to a number of people without, as it were, being filled with genetic matter — form is perceived visually on its own, by itself. The other is that this form is not genetically individualized, specialized, but is the universal, archetypal form of man. We may be reminded here of Goethe's archetypal plant. In the Resurrection, it was not the individually unique body of Jesus of Nazareth, but the universal archetype of Man that became visible to his disciples.

Therefore, the disciples and Mary Magdalene who had known Jesus so well could not recognize him in the Risen Christ, because he had shed matter as well as any genetic distinction in raising the generally spiritual human form. In the Resurrection, something entered into visibility, was made manifest, experienced by contemporaries, which had become forgotten and hidden in the course of the history of mankind.

We may now understand the importance of the story of the twofold conception in the Gospels of Matthew and Luke and the over-riding stress on the non-genetic conception through the Holy Spirit. Only when we are helped to grasp the distinction between the form-principle and the genetic principle can we hope to comprehend the mystery of the Resurrection.

Just as the conception of Jesus which for the first time revealed that the duality of universal form and individual heredity is equally valid for each of us, in fact inherent in our biological make-up, so is Christ's Resurrection, although truly unique, valid for each of us as a potential, a promise.

While the recognition of the form-conception through the Holy Spirit opens up a new understanding of the phenomenon of the Resurrection, it does not by itself explain the raising of the spiritual-physical human form. It is to the account of the second descent of the Holy Spirit at the Baptism of Jesus in the Jordan that we have to turn here. In the Gospels of Mark and John, this event in the life of Jesus is related in the first chapters. Matthew and Luke bring it towards the end of their third chapters — Luke just before and directly leading over to the enumeration of the genealogy of Jesus, thus contrasting the Baptism as a spiritual conception with the genealogical one. Here the appearance of the descending dove indicates that the cosmic sun-being of Christ as the creating force in the zodiac unites the universal archetype of the form of man with the bodily incarnation of the individual man, Jesus. As we have maintained, it is this archetypal human form that is raised in the Resurrection of Christ, the spiritual-physical, non-material, form of Man. Of course, this is only one of the many aspects of the significance of the union of the Christ Logos with Jesus of Nazareth.

Meanwhile, most of us believe that our form, and form altogether, exists only as the shape of matter. Once we have understood that form has an independent existence, we can set out towards the direct experience and certainty of our own indestructible, undying, eternal being. For the present, we hold that man is 'really', 'objectively' the end-product of a long, long animal evolution. This is undoubtedly true in its application to our genetic inheritance. The theory of evolution has proved to be an extraordinarily elucidating hypothesis, but it has had destructive consequences for our self-understanding, as we have failed to recognize the specific limitations of its relevance. It has made us believe that scientific evidence is incompatible with religion, and has caused us to be blind to the philosophical and spiritual implications of modern scientific insights. Schools and popular scientific literature have one-sidedly and nearly exclusively propagated a knowledge of the genetic element in biology, suppressing the other half, the development of form. A similar situation has developed in the context of modern physics with possibly even more dire results.

10. Form and individuality

We want now to take up the question of the primacy of the human form in its cosmic origin and its link to the individuality or ego in man. At the beginning of Chapter 6 we referred to P. Weiss's observations of surface patterns in living organisms which often did not coincide with cellular or molecular boundaries. Similarily, we see that door and window surrounds of houses are created out of whole and half bricks, which assures us that houses do not spontaneously arise out of self-regulating forces inherent in the bricks. Of course, the erection of a brick house presupposes an architect or builder, a creator. The same could be assumed in relation to man but the idea tends to be an anathema in our time. This is probably largely attributable to the unfortunate misconception that sees such a creator in the likeness of man, in the image of a large father-figure. In Chapter 6, we cited many other authors' contributions all of which supported the independence of the development of form. But what arguments can we bring to respond to the massive evidence of palaeontology which clearly shows that higher forms of animals appeared later than the rudimentary ones, and man last of all? There seems to be no doubt in the validity of these findings, proving that the physical-material form of man is a late, relatively more recent phenomenon and seems to be part of a slow, gradual genetic evolution. What alternative evidence could we possibly offer to support the statement that the development of form is something in itself?

The first is, of course, an argument of meaning and definition. We do not mean with form a quality of matter, but an element in itself that may or may not become manifest in matter. The plan of a house precedes its actual erection, but again, the house itself

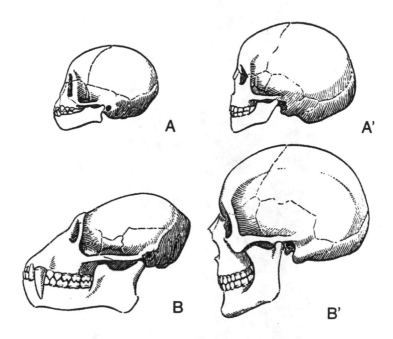

Figure 21. Comparison of juvenile and adult skulls of gibbon (A, B), and of human (A', B').

may never materialize. But what observable evidence can we put forward in support of our hypothesis?

Figure 21 shows four skulls — two small ones (A and A[1]) which look fairly similar, and two larger ones, one (B) of which is obviously an animal skull, the other (B[1]) equally obviously human. The two smaller, fairly similar skulls are the embryonic forms of full-term animal and human skulls respectively. Portmann illustrates a general principle here, namely that the human form is *primary* to the animal form, so that young and embryonic animals are still similar to the human form and achieve their specific animal form only gradually as they develop.

This is even more powerfully illustrated by a photograph of a rhesus monkey embryo taken surgically from its mother's womb, when compared with an adult rhesus monkey (Figures 22 and 23). The embryo appears like an old man, straight and upright with

137

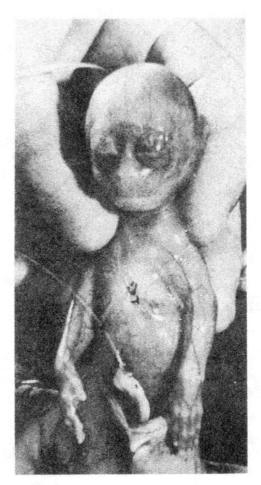

Figure 22. Foetus of rhesus monkey (104 days).

domed forehead and beautifully shaped hands and fingers, whereas the adult animal has a fur covering right up to the face, no forehead, furry paws and a long tail.

Could these phenomena of ape and monkey development indicate a general principal that the original human form becomes one-sidedly specialized in animal, or at least in vertebrate development, in materializing as matter? This idea has, in fact, been delightfully documented by W. Schad (1982), who presents many examples in support of it.

Figure 23. Adult rhesus monkey.

When modelling the human head or skull in clay with students, I have often observed the following: when one of them had clay that was slightly too wet and accidently dropped his model on the floor, he would pick up, to the great amusement of the group — an ape's skull. I even developed this as an exercise with the students, demonstrating that dropping a model of the human skull could result in an ape, even horse, dog or other animal likeness, but never in a human one.

It appears that gravitational impact pushes the form of the

139

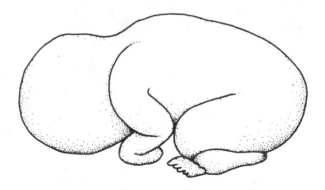

Figure 24. Lateral view of human foetus at term.

human head towards the animal head-form, but never the other way around. Does this hold good for the body-form as well? While it seems more obvious in connection with the head, it is also applicable to other parts of the body. It is easier to imagine our human hand becoming a mole's paw by closing and slightly bending fingers or a horse's hoof by enlarging the middle finger and nail and retracting the other four fingers, or again a cat's paw, or a duck's foot, rather than the reverse exercise in which all these could possibly become a human hand in all its refinement.

The head, however, seems to occupy a special, preferred place in the human form. Towards the end of Chapter 8 was described that what appears to be the first emergence of the embryonic body with its head and limbs around the fifth week, is really only the beginning of the head-form, and that what looks like the limbs is really the outline of what is becoming the upper and lower jaws. In addition, there is an astonishing similarity in the form of the foetus in the womb before birth and in the human brain (Figures 24 and 25). The head corresponds to the frontal pole of the brain, the back to the parietal lobe, the buttocks to the occipital lobe, the knees to the temporal lobes, the feet to the cerebellum and the navel cord to the spinal cord. It is only at birth that this brain-form unfolds once more into the human form. We are not only normally born head first, but are born *out of the form of the head.*

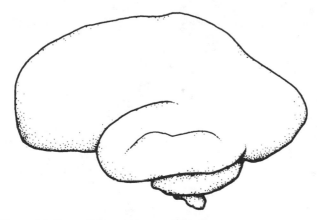

Figure 25. Lateral view of brain of human foetus at term.

In his lectures on reincarnation, Rudolf Steiner describes a metamorphosis from body-form to head-form from one incarnation to the next. What was body in the past becomes head in the future. It is as though we shed the form of our head in death, metamorphosing our body-form into future head-form so that we incarnate through birth with the head-form we bring with us from the past, developing out of it our new body-form. This metamorphosis can readily be observed in the human skeleton (Figure 26).

If we imagine the skeleton without head, kneeling, arching backwards, the spine compressed and foreshortened into the three vertebrae forming the base of the skull, we see belly and chest arching over it as the skull-cap, the arms as maxillae hanging down, angled at the elbows which form the cheekbones, the hands above the feet. The fingers and toes indicate the 20 milk teeth, the sixteen wrist-and-ankle bones the 32 second teeth. The knee is the angle of the mandible; the hip joint, the joint of the lower jaw. The hollows between clavicles and shoulder appear as the eye sockets.

There is a strange flaw in this transformation, however. While the hands in supination fit well into their role as upper jaws, the feet which are fixed in pronation, hold their heel — the chin — upwards and the toes, hence teeth, downwards. But when we look at the lower arm and leg in the skeleton, we see that when radius and ulna are parallel and not crossed, the hand is supinated,

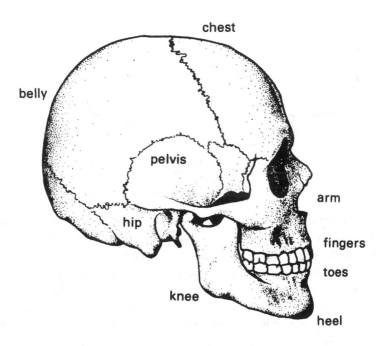

Figure 26. The human skull seen as transformation of the body.

while in the leg where fibula and tibia cannot cross at all, the foot is permanently pronated, which fact causes our problem in the metamorphosis we postulate.

In addition, we observe that radius and ulna are tapered, whereas fibula and tibia are straight, the tibia being equally thick, the fibula equally thin throughout their entire respective lengths, while the radius is thin proximally and thick distally, and the ulna the reverse. When ulna and radius are crossed, with the hand pronated (Figure 27), one can see the proximal part of the ulna and the distal half of the radius as the tibia, and the proximal half of the radius and the distal half of the ulna as the fibula — as if the bones had been broken in pronation and reunited crossways. This could possibly have happened during the transition from an

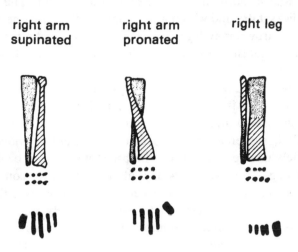

Figure 27. Tibia shown as being half radius and half ulna, and fibula as half ulna and half radius.

aquatic to a terrestrial existence, but I have not been able to find definitive evidence for this.

The Gospel of John, however, reports that it was the custom at the time to break the legs of the victims of crucifixion to speed death, but that in fulfilment of Scripture, Christ's legs were not broken. This may well point to the significance of an originally unbroken form of leg that could be supinated like the hand.

The metamorphosis from body to head has been beautifully described and illustrated by L. F. C. Mees (1984). That we start as head and grow in time towards body is amply borne out by the facts of child development. The head of the embryo at eight to nine weeks is nearly half the size and weight of the whole body. In the new-born child, the head is only a quarter of the child's size; in the adult, it is one seventh. Not only in the size and features of body-development, but also in the development and unfolding of the child's personality may we regard the head as the original initiator.

We usually state that children 'grow up', but particularly in the very young child, we can observe that this is misleading. When a calf or foal is born, it is dropped to the ground. The mother animal licks it dry and within a few minutes, the little one wriggles on to its sturdy, lanky legs, lifts its small head and after pushing vigorously against the mother's udder, begins to drink — it has 'grown up'.

We shall never witness such a thing in the new born human infant. Not only is a human infant not dropped; it reaches the floor usually only after the first year of life. The human infant is embedded, suspended in an ocean of involuntary movements, completely unable to control its limbs. Motor-control begins with the focusing of the eyes a few days after birth. Motor-control then descends from the head to the neck; by three weeks, the child can lift its head and respond to a stimulus. By three months, motor-control has descended into the arms and hands — the child can grasp objects seen and chosen. By five or six months, the child sits up, voluntary control having descended to the hips. By eight or nine months, the child can crawl, control having reached the knees, and by the end of the first year, the child has reached the earth with its feet; it stands and walks. Thus we can see movement control *descending* from the head to the feet in the young infant. I have described this descent in some detail in my book *Children in Need of Special Care*.

I have referred to these aspects of head-body relationships in an attempt to foster, support and strengthen a direct experience of relating to the original zodiacal Man, the Adam Kadmon, the human form expanded into the circumference and width of the whole solar system, as I described earlier. In moments in which the Adam Kadmon, the infinitely expanded human form, becomes an experiential certainty to a person, he enters a different form of consciousness, an expanded, peripheric consciousness, while our normal, day-to-day consciousness is a centred and circumscribed one.

In an expanded, peripheric consciousness, one feels at one with the world — the world and my ego are one and the same. My ordinary, normal experience of the world only arises to the extent to which I have withdrawn my ego from the world and make the world my non-ego, my ego the non-world.

This becomes more readily comprehensible when we follow up

child development from the first to the second (and third) year, not only in relation to the development of movement, but also to an infant's mode of experience and consciousness. In its first months, an infant experiences no objects. All experience is self-related; whatever is experienced is the infant's, in what has been described as a state of 'omnipotence'. Only step by step does the infant's initially expanded, diffuse ego begin to centre and withdraw inward from the world which now gradually begins to appear as a world of objects. The development of experience and consciousness goes parallel with the development of movement in the first year, leading to the development of language in the second when the newly perceived objects acquire their names.

Up to the Second World War, it took three years before children began to experience themselves as ego and referred to themselves with the personal pronoun 'I'. Since then, this mysterious event in a child's life has advanced by a whole year. Nearly all healthy children now refer to themselves as 'I' by the end of their second year.

As far as I know, this phenomenon has not previously been noted or described, yet I believe it to be of fundamental importance both for the self-understanding of man and for the pathologies of child development. I want to turn here only to the first of these two aspects.

A child learns his mother tongue by hearing and then imitating the spoken word. (Only baby-language is universal and genetically based). The word 'I' is the only exception. No one has ever addressed a child as 'I', and at the age of two, a child's thought-processes are not so developed as to draw the conclusion that if all others call themselves 'I' he should refer to himself as 'I', as well. The experience of oneself as 'I' is probably the only spiritual experience that every ordinary person has in infancy. It occurs at a time when the child goes through the profoundly shaming phase of toilet training. When we openly and with loving interest observe children who are going through this phase, which Erikson describes as the stage of 'autonomy versus shame', we are impressed by the daring impossibility of human existence: the biological, natural creature-bondage in which lights up the brilliant spark of the divine ego, the dewdrop of the totality of the cosmic divine powers. We often see children of that age being wheeled in their prams, riding like little kings in a carriage, looking at us adults with the gaze of a detached, divine emperor, not certain whether we are also of the nature of ego.

I am aware of my bias in favour of the young child, but I know I share this not only with most child psychiatrists, but with very many people. I often believe one encounters in such young children a hope and a promise of making human existence divine; of achieving the impossible task of giving full expression to the creator principle in the nature of the creature.

In these last pages, I have tried to link the human form to the 'I'-experience. I draw justification for this attempt from Rudolf Steiner's spiritual characterization in his *Occult Science* of the human 'I' or ego as being a deed, a gift of the hierarchical Spirits of Form, the Exousiai. I believe this can also be found in Goethe's scientific work, summed up charmingly in his little verse:

> Vom Vater hab' ich die Statur
> Des Lebens ernstes Führen.
> Vom Mütterchen die Frohnatur,
> Und Lust zum Fabulieren.
>
> (From my father I received my from,
> My dignified and earnest state.
> From my mother my ebullient nature,
> And tendency to fabulate.)

It links form and 'I' or ego to the father principle; sentient and life or vegetative nature to the mother principle. And yet one cannot possibly speak of an identity of form and ego without referring to the profound polarity between the two. Our deliberations have sprung from the contemporary biological discovery of the independence, the separateness of form development in living organisms from their genetic development. We stressed that *form* was what united all men, what we all had in common, what we shared, and that our genetic make-up was responsible only for our individual constitutional differentiations.

How can our ego, the core and centre of our individuality, be linked to our form rather than to the individualizing and differentiating genetic element? One could, of course, push the problem on to the word 'individual', which stems from indivisible, inseparable; and one could say that genetic traits separate, divide humanity and all living organisms. One could also say that form as such is indivisible, while matter and individual organisms are divisible. But we cannot be satisfied and acquiesce with this argument alone. After all, we all share the one human *form*, but we certainly do not share one ego. In fact, we are quite acutely and predominantly aware of the separate, divisive nature of our ego. No one else is 'I'. No one else can possibly be 'I'. The whole world only exists around me because it is not 'I'. Our present, ordinary, personal ego-experience isolates us, separates us, divides us from one another, makes us *solitary*. Does that necessitate our continuing urge to seek fusion with other individuals in sexual love, to flee from, to transcend the isolating power of our singular ego experience?

It is, however, not a spiritual ego-experience that draws into the two-year-old child. It is a corrupted, defamed ego-experience of the ordinary adult of today who is convinced that in all honesty, we have to accept that we are an animal species, in some ways well-adapted, but determined by our genes, part of inevitable binding natural laws.

The importance of the biological discovery of the independence of form lies in the fact that it can lead us to this other experience of ego that we have all had in our infancy and which can freely reach into adulthood if we so decide; the expanded ego that does not isolate, separate, divide, but that embraces, includes all within it.

We can achieve this only in moments, as we must ever return to our ordinary day-centred consciousness, but we then bring with us in this return, a Christ-experience that gives us a certainty of being one-with-the-world, of being sustained by the world, of belonging and of being ultimately responsible, not only for all living beings, but for the earth as a whole.

The events of Pentecost as they are described in the Acts of the Apostles when the Holy Spirit again descended, filling the disciples so that they began to speak in other tongues, must have been an occurrence of expanded ego-consciousness. Suddenly, the separating barriers of ordinary isolating ego-experience were overcome and filled with the universal, expanded ego-experience; the disciples could be understood by all who were present in Jerusalem. They then had to return to their customary ways, but retained the imprint of a new faith.

At the beginning of Chapter 7 we mentioned that the relevant discoveries in biology had been preceded by discoveries in physics which transcended the Newtonian natural laws and their under⁊ lying principles. Again it was only the technically applicable aspects of the discoveries that were followed up and developed and which led to the threat of a nuclear holocaust, while the philosophical and spiritual implications were largely neglected.

I should like here to take up one of the relatively simpler concepts of modern physics which, like the experience of the independence of form from matter, should have become generally thinkable and accepted. This is the idea that the linear one-directional image or experience of time as separate from space changes, becomes untenable, and has to be replaced by the idea of time as a fourth dimension of space, if we want to enter into sub-atomic or stellar dimensions. Like infinity to which we referred in Chapter 8 the concept of a fourth dimension is easily followed mathematically, but difficult to imagine. We arrive at the second dimension by making on a line of, say, two units' length, a square of four square units, multiplying 2×2; and at the third dimension by constructing a cube of 8 cubic units by multiplying $2 \times 2 \times 2$. If we continue to $2 \times 2 \times 2 \times 2 = 16$, we have reached the fourth dimension, but find that we cannot form any spatial image of it. We may then sense that rather than develop nuclear weapons, we might have gained more by trying to understand what the consequences of four dimensional experience might be.

We may use a comparison to help our imagination which like most comparisons does not completely stand up to logical criticism, but is nonetheless helpful. We wish to know what it would be like to live in the awareness of a fourth dimension. We could approach this task by stating that the experience of a fourth dimensional space to us, who are used only to three-dimensional experience, may be comparable to the experience of a being used only to two dimensions within our three-dimensional space. Let us imagine ourselves as a two-dimensional creature, a dot in a two-dimensional world — the surface of this page. On this page, we have drawn two lines meeting at an angle of, say, 60°, but well away from the creature. The creature perceives these lines as one straight line, but as he moves towards this 'straight' line, the latter suddenly begins to come towards him from the left and the right, seeming to close in upon him, causing him to feel attacked and trapped; something unforeseen is happening to him.

We, however, from our three-dimensional vantage point know that what to the creature is unforeseeable future has become experience to him only as the result of his own action, his movement towards the angular form. If we could learn to experience in four dimensions, we might realize that our own unforeseeable future is ever-present, is a structure pre-determined and pre-existent, already there and yet always becoming our destiny in consequence of our own action and initiative. Would this be so? Would we learn to experience that what happens to us is ours, is brought about by us and yet is always there? What consequences would such experience have in relation to our anxieties and uncertainties that make us invest most of our resources in insurance and precautions? What changes, what growth in our self-experience could arise if we would take earnestly the philosophical aspects of modern scientific development and cultivate a consciousness of four-dimensional experience?

Bibliography

Adams, George. 1965. *Strahlende Weltgestaltung*. Dornach: Philosophisch-Anthroposophisch.

Appenzeller, Kaspar. 1976. *Die Genesis im Lichte der menschlichen Embryonalentwickelung*. Basel: Zbinden.

Ayala, F. J., & T. Dobzhansky (editors). 1974. *Problems of Reduction in Biology. Studies on the Philosophy of Biology*. London: Macmillan.

Bateson, William. 1909. *Mendel's Principles of Heredity*. Cambridge: University Press.

Beer, G. R. de, *see* Huxley, Julian S. & G. R. de Beer.

Blackler, A. *see* Fishberg, M. & A. Blackler.

Blechschmidt, Erich. 1970. *Vom Ei zum Embryo*, Hamburg: Rowohlt.

Boyd, J. D. *see* Hamilton, W. J., J. D. Boyd, & H. W. Mossman.

Campbell, Joseph. 1973. *The Masks of God*. London: Paladin.

Capra, Fritjof. 1976. *The Tao of Physics*. London: Fontana.

Cockerell, Sidney Carlyle. 1969. *Old Testament Miniatures*. London: Phaidon.

Colegrave, S. 1979. *The Spirit of the Valley*. London: Virago.

Colum, Padraic. 1930. *Orpheus: Myths of the World*. New York: Macmillan.

Corliss, C. E. 1976. *Patten's Human Embryology*. New York: McGraw Hill.

D'Arcy Thompson, W. *see* Thompson, W. D'Arcy.

Darwin, Charles R. 1859. *On the Origin of Species*. (Reprint 1968 Harmondsworth: Penguin.)

Dawkins, Richard. 1978. *The Selfish Gene*. London: Paladin.

Dowdeswell, William H. 1963. 3 ed. *The Mechanism of Evolution*. London: Heinemann.

Driesch, Hans. 1891. *Die mathematisch-mechanische Betrachtung morphologischer Probleme der Biologie*. Jena.

Edwards, Lawrence. 1982. *The Field of Form*. Edinburgh: Floris.

Eliade, Mircéa. 1968. *Myths, Dreams and Mysteries*. London: Fontana.

Erikson, E. H. 1963, 2 ed. *Childhood and Society*. New York: Norton. (Reprint 1965 Harmondsworth: Penguin.)

Fishberg, M., & A. Blackler, 1961. 'How Cells Specialize'. *Scientific American*. September.

Gauquelin, Michel. 1973. *The Cosmic Clocks*. St Albans: Paladin.

Gebser, J. 1965. *Abendländische Wandlung*. Frankfurt: Ullstein.

——. 1974. *Verfall und Teilhabe*. Salzburg: Otto Muller.

Goethe, Johann Wolfgang. *Zur Naturwissenschaft*.

Goodwin, Brian. 1976. 'On Some Relationships between Embryogenesis and Cognition'. *Theoria to Theory*. 10:33–34.

——. 1979. 'On Morphogenetic Fields'. *Theoria to Theory*. 13:109–14.

Gray, G. W. 1957. 'The Organizer'. *Scientific American*. November.

Grimal, Pierre. 1965. Larousse World Mythology. London: Hamlyn.

——. 1967. *Mythen der Völker*. Frankfurt: Fischer.

Hamilton, William James, J. D. Boyd, & H. W. Mossman. 1945. *Human Embryology*. Cambridge: Heffer.

Hamilton, William James & H. W. Mossman. 1972. 4 edn. *Hamilton, Boyd and Mossman's Human Embryology*. Cambridge: Heffer, and Baltimore: Williams & Wilkins.

Hoerner, Wilhelm. 1978. *Zeit und Rhythmus*. Stuttgart: Urachhaus.

Hoffmeister, M. 1979. *Die übersinnliche Vorbereitung der Inkarnation*. Basel: Pforte.

Huxley, Julian S. & G. R. de Beer. 1934. *The Elements of Experimental Embryology*. Cambridge: University Press.

Julius, F. H. 1956. *Die Bildersprache des Tierkreises*. Stuttgart: Mellinger.

Jung, Carl Gustav. 1964. *Man and his symbols*. London: Aldus.

Kerényi, Károly. 1966. *Die Mythologie der Griechen*. München: Deutscher Taschenbuch.

Koestler, Arthur, & J. R. Smithies (Editors). 1969. *Beyond Reductionism*. London: Hutchinson.

König, Karl. 1927a. 'Menschliche Embryonalentwickelung. Ontogonie und Phylogenie'. *Gaie Sophie* 2. Dornach.

——. 1927b. 'Einige geisteswissenschaftliche Betrachtungen über die Eihüllen und die erste Anlage des Menschenkeimes'. *Natura* 1.327–31.

——. 1927c. 'Über die Grundkräfte welche den menschlichen Embryo gestalten.' *Natura* 2.108–18.

——. 1928. 'Der Ursprung des Menschen in seiner Beziehung zur menschlichen Embryonalzeit.' *Natura* 3.24–36.

——. 1967. *Embryologie und Weltentstehung*. Freiburg: Kommenden.

——. 1984. 4 ed. *The First Three Years of the Child*. New York: Anthroposophic, and Edinburgh: Floris.

Laing, Ronald D. 1967. *The Politics of Experience and the Bird of Paradise*. Harmondsworth: Penguin.

Lamarck, Jean Baptiste Monet de. 1809. *Philosophie zoologique*. Paris.

BIBLIOGRAPHY

Langman, Jan. 1969. *Medical Embryology*. Baltimore: William & Wilkins.

Malthus, Thomas R. 1798. *An Essay on the Principle of Population*. (Reprinted 1970. Harmondsworth: Penguin.)

Mees, L. F. C. 1984. *Secrets of the Skeleton*. New York: Anthroposophic.

Mendel, Gregor, 1866. *Über Pflanzen-Hybriden*. Bruenn.

Monod, Jacques. 1972. *Chance and Necessity*. London: Collins.

Moody, Raymond. 1977. *Life after Life*. London: Corgi.

Mossman, H. W. *see* Hamilton, W. J., (J. D. Boyd), & H. W. Mossman.

Needham, N. Joseph. 1934. *A History of Embryology*. Cambridge: University Press.

Nilsson. Lennart. 1967. *Ein Kind Entsteht*. Gütersloh: Bertelsmann.

——. 1967. *The Everyday Miracle: A Child is Born*. London: Lane.

Pai, Anna C. 1974. *Foundations of Genetics*. New York: McGraw Hill.

Popper, Karl. 1976. *Unended Quest*. London: Fontana.

Portmann, Adolf. 1948. *Einführung in die vergleichende Morphologie der Wirbeltiere*. Basel: Schwabe.

Schad, Wolfgang. 1982. *Die Vorgeburtlichkeit des Menschen*. Stuttgart: Urachhaus.

Schell, Jonathan. 1982. *The Fate of the Earth*. London: Picador.

Sheldrake, Rupert. 1981. *A New Science of Life*. London: Blond & Briggs.

Spemann, Hans. 1938. *Embryonic Development and Induction*. New Haven: University Press.

Steiner, Rudolf. 1883–97. *Goethes naturwissenschaftliche Schriften*. (Reprinted 1973. Dornach: Steiner. GA 1.)

——. 1897. *Goethes Weltanschauung*. (Reprinted Dornach: Steiner. GA 6.)

——. 1966. *Man's Being, his Destiny and World Evolution*. New York: Anthroposophic.

——. 1972. *Building Stones for an Understanding of the Mystery of Golgotha*. London: Steiner.

——. 1975. *Between Death and Rebirth*. London: Steiner.

——. 1979. *Occult Science, an Outline*. London: Steiner

——. 1983. *An Occult Physiology*. London: Steiner.

Stevens, A. 1982. *Archetype*. London: Routledge & Kegan Paul.

Sturtevant, Alfred H. 1965. *A History of Genetics. New York: Harper & Row.*

Thompson, W. D'Arcy. 1942. On Growth and Form. Cambridge: University Press.

Waddington, C. H. 1953. 'How do Cells Differentiate?' *Scientific American*. September.

Weihs, Thomas J. 1971. *Children in Need of Special Care*. London: Souvenir.

Whicher, Olive. 1971. *Projective Geometry*. London: Steiner.

Wilmar, Frits. 1979, *Vorgeburtliche Menschwerdung*. Stuttgart: Mellinger.

Winnicot, Donald W. 1965. *The Family and Individual Development*. London: Tavistock.

Zolla, Elemire. 1981. *The Androgyne*. London: Thames & Hudson.

Index

INDEX

For news on all our **latest books**,
and to receive **exclusive discounts**,
join our mailing list at:

florisbooks.co.uk

Plus subscribers get a FREE book
with every online order!

We will never pass your details to anyone else.

CPSIA information can be obtained
at www.ICGtesting.com
Printed in the USA
BVHW090917211220
596164BV00016B/433

9 781782 504993